Nature's Numbers

Expanding the National Economic Accounts
to Include the Environment

William D. Nordhaus and Edward C. Kokkelenberg, *Editors*

Panel on Integrated Environmental and Economic Accounting

Committee on National Statistics

Commission on Behavioral and Social Sciences and Education

National Research Council

NATIONAL ACADEMY PRESS
Washington, D.C.

NATIONAL ACADEMY PRESS • 2101 Constitution Avenue, N.W. • Washington, DC 20418

NOTICE: The project that is the subject of this report was approved by the Governing Board of the National Research Council, whose members are drawn from the councils of the National Academy of Sciences, the National Academy of Engineering, and the Institute of Medicine. The members of the committee responsible for the report were chosen for their special competences and with regard for appropriate balance.

Support of the work of the Committee on National Statistics is provided by a consortium of federal agencies through a grant from the National Science Foundation (Number SBR-9709489). The project that is the subject of this report is supported by the Bureau of Economic Analysis of the U.S. Department of Commerce through Grant No. SBR-9409570 between the National Academy of Sciences and the National Science Foundation. Any opinions, findings, conclusions, or recommendations expressed in this publication are those of the author(s) and do not necessarily reflect the view of the organizations or agencies that provided support for this project.

Additional copies of this report are available from National Academy Press, 2101 Constitution Avenue, N.W., Washington, D.C. 20418
Call (800) 624-6242 or (202) 334-3313 (in the Washington metropolitan area)
This report is also available on line at **http://www.nap.edu**

Printed in the United States of America

iii

Acknowledgments

The Panel on Integrated Environmental and Economic Accounting wishes to thank the many people who contributed to its work.

The project was sponsored by the Bureau of Economic Analysis of the U.S. Department of Commerce. At the beginning of our work, and throughout the deliberations, the panel was helped by the director of the bureau, J. Steven Landefeld, as well as its staff, who provided background on their work on the U.S. Integrated Environmental and Economic Satellite Accounts and discussed the issues involved in developing the accounts. We particularly thank Gerald F. Donahoe, Bruce T. Grimm, Arnold J. Katz, Stephanie H. McCulla, Robert P. Parker, and Timothy Slaper for their help in explaining the complexities of environmental accounting.

The panel's work also benefited from the contributions of Graham Davis of the Colorado School of Mines, who helped us incorporate the economics of subsoil assets, and James Hrubovcak of the U.S. Department of Agriculture (USDA), who was helpful in sharpening our understanding of water and air issues. In addition, the panel benefited from the participation of Peter Feather and Daniel Hellerstein, both of the USDA, in our discussions of nonmarket valuation.

We are also grateful for the perspectives of several experts who made major presentations to, or held discussions with, the panel, including Gerald Gravel of Statistics Canada, Kirk Hamilton of the World Bank, John Hartwick of Queens University in Ontario, Peter Bartelmus of the United Nations, Richard Haines of the U.S. Forest Service, and Craig Schiffries of the National Research Council.

We also note that the many members of the London Group of national income accountants, who are concerned with integrated environmental and economic accounting, allowed us to attend their annual meetings and shared many documents with us, thus giving the panel a broader perspective concerning the problems we considered. We are particularly thankful for insights provided by Ann Harrison of the Organization for Economic Cooperation and Development and Henry Neuburger of Her Majesty's Government, United Kingdom.

This report, the collective product of the entire panel, reflects the dedication and commitment of its individual members. All of the panel members participated in many meetings and discussions and in reviewing drafts and contributing sections to the final report. In addition, John Tilton led a subpanel on minerals and John Reilly and Henry Peskin led a subpanel on renewable and environmental resources. Clark Binkley was particularly helpful in developing the sections of the report on forestry, and Martin Weitzman was instrumental in developing the material on sustainability.

The panel was extraordinarily lucky to have the assistance of Edward Kokkelenberg, the study director, who had responsibility for organizing and coordinating panel and subpanel meetings, gathering much of the written material, attending the London Group conference, arranging for consultants, and preparing the report. Without his skills and dedication, the report could not have been produced in the time available.

The panel was established under the auspices of the Committee on National Statistics. Miron Straf, director of the committee, was instrumental in developing the study and providing guidance and support to the panel and staff. The committee, under the chair first of Norman Bradburn and later of John Rolph, had the responsibility for establishing the panel and monitoring its progress. Deputy director Andrew White helped us in the final stages to develop sharp recommendations and navigate the requirements of the National Research Council.

Other members of the staff included Joshua Dick, Cassandra Shedd, Jennifer Thompson, and Anu Das; they provided excellent administrative, editorial, and research support for the study and the report. We also thank Rona Briere, who helped us improve the report through technical editing. To all we are most grateful.

Our report has been reviewed in draft form by individuals chosen for their diverse perspectives and technical expertise, in accordance with procedures approved by the National Research Council's Report Review Committee. The purpose of this independent review is to provide candid and critical comments that will assist the institution in making the published report as sound as possible and to ensure that the report meets institutional standards for objectivity, evidence, and responsiveness to the study charge. The review comments and draft manuscript remain confidential to protect the integrity of the deliberative process.

We wish to thank the following individuals for their participation in the review of this report: Theodore W. Anderson, Department of Statistics, Stanford University (emeritus); Kenneth J. Arrow, Department of Economics, Stanford University; Peter Bartelmus, Statistics Division, United Nations; James R. Craig, Geologic Sciences, Virginia Polytechnic Institute and State University; Martin H. David, Department of Economics, University of Wisconsin; Michael R. Dove, School of Forestry and Environmental Studies, Yale University; Theodore R. Eck, AMOCO, Chicago, IL; Charles Hulten, Department of Economics, University of Maryland; Daniel M. Kammen, Woodrow Wilson School of Public and International Affairs, Princeton University; Arthur H. Lachenbruch, U.S. Geologic Survey, Menlo Park, CA; Thomas A. Louis, School of Public Health, University of Minnesota; Donald Ludwig, University of British Columbia (emeritus); Thomas C. Schelling, School of Public Affairs, University of Maryland; Burton H. Singer, Office of Population Research, Princeton University; and Robert M. Solow, Department of Economics, Massachusetts Institute of Technology.

Although the individuals listed above have provided constructive comments and suggestions, it must be emphasized that responsibility for the final content of this report rests entirely with the panel and the National Research Council.

This report and its many antecedents over the last two decades owe their existence, high quality, and purpose to the pioneering work of the late Robert Eisner of Northwestern University. Professor Eisner was a member of the panel and gave us his wisdom and guidance throughout our deliberations. Bob Eisner died in November 1998 after the report was completed. I speak for the panel in saluting his many contributions; we will miss him.

William D. Nordhaus, *Chair*
Panel on Integrated Environmental
and Economic Accounting

Contents

ix

Nature's Numbers

Executive Summary

 This report addresses the question of whether the U.S. National Income and Product Accounts (NIPA) should be broadened to include activities involving natural resources and the environment. The NIPA are the most important measures of overall economic activity for a nation. They measure the total income and output of the nation; their purpose is to provide a coherent and comprehensive picture of the nation's economy.

A central principle underlying the national accounts is to measure production and income that arise primarily from the market economy. However, the NIPA's focus on market activities has raised concerns that the accounts are incomplete and misleading because they omit important nonmarket activities, such as nonmarket work, the services of the environment, and human capital. In response to these concerns about standard measures of economic activity, private scholars and governments have endeavored to broaden the national accounts in many directions. Most recently, attention has focused on extending the accounts to include natural resources and the environment. The guiding principle in extended national accounts is to measure as much economic activity as is feasible, whether that activity takes place inside or outside the boundaries of the marketplace.

Intensive work on environmental accounting began in the Bureau of Economic Analysis (BEA) of the U.S. Department of Commerce in 1992. Shortly after the first publication of the U.S. Integrated Environmental and Economic Satellite Accounts (IEESA) in 1994, Congress directed the

Commerce Department to suspend further work in this area and to obtain an external review of environmental accounting. A panel working under the aegis of the National Research Council's Committee on National Statistics was charged to "examine the objectivity, methodology, and application of integrated environmental and economic accounting in the context of broadening the national accounts" and to review "the proposed revisions . . . to broaden the national accounts." This report presents the panel's findings and recommendations.

INTEGRATED ENVIRONMENTAL AND ECONOMIC ACCOUNTING AND ITS BENEFITS TO THE NATION

BEA developed the IEESA because of the growing importance of environmental accounting both in the United States and abroad. Better natural-resource and environmental accounts have many benefits. They provide valuable information on the interaction between the environment and the economy; help in determining whether the nation is using its stocks of natural resources and environmental assets in a sustainable manner; and provide information on the implications of different regulations, taxes, and consumption patterns.

More generally, augmented NIPA that encompass market and nonmarket environmental assets and production activities would be an important component of the U.S. statistical system, providing useful data on resource trends. The rationale for augmented accounts is solidly grounded in mainstream economic analysis. BEA's activities in developing the environmental accounts are consistent with an extensive domestic and international effort both to improve and to extend the NIPA.

The panel concludes that extending the U.S. national income and product accounts (NIPA) to include assets and production activities associated with natural resources and the environment is an important goal. Environmental and natural-resource accounts would provide useful data on resource trends and help governments, businesses, and individuals better plan their economic activities and investments. The rationale for augmented accounts is solidly grounded in mainstream economic analysis. BEA's activities in developing environmental accounts (IEESA) are consistent with an extensive domestic and international ef-

---

[1]Paragraphs in boldface in this executive summary reflect recommendations in the main report. The numbers after each paragraph refer to the corresponding recommendations in the chapters that follow; for example, Recommendation 5.1 is the first recommendation in Chapter 5.

fort to both improve and extend the NIPA. (Recommendation 5.1)[1]

There are two possible approaches to developing nonmarket and environmental accounts: a phased and a comprehensive approach. BEA's proposal for developing the IEESA envisions use of the phased approach, adding satellite accounts for natural-resource and environmental assets in three phases—starting with subsoil mineral assets, expanding to renewable and other natural resources such as timber in forests, and only then addressing nonmarket environmental assets such as clean air and water. Under the comprehensive approach, a broad set of nonmarket accounts would be developed in parallel with the near-market accounts. BEA would develop accounts not only for the minerals and near-market sectors, but also for nonmarket activities and assets.

If the phased approach is undertaken, a useful initial step would be to refine the initial estimates of subsoil minerals. Constructing forest accounts, focusing initially on timber, is a natural next step for integrated environmental and economic accounts. Other sectors that should be high on the priority list are those associated with agricultural assets, fisheries, and water resources.

Although recognizing the value of the phased approach, the panel finds that developing comprehensive nonmarket accounts is of the greatest substantive importance for augmented accounting and for policy purposes. The panel does not, however, underestimate the challenges involved in developing nonmarket accounts. The process will require resolving major conceptual issues, developing appropriate physical measures, and valuing the relevant flows and stocks.

The panel concludes that developing a set of comprehensive nonmarket economic accounts is a high priority for the nation. Developing nonmarket accounts to address such concerns as environmental impacts, the value of nonmarket natural resources, the value of nonmarket work, the value of investments in human capital, and the uses of people's time would illuminate a wide variety of issues concerning the economic state of the nation. (Recommendation 5.2)

At present, BEA does not plan to redefine the core NIPA to include flows or investments in natural resources and the environment. Natural-resource and environmental flows will instead be recorded in satellite or supplemental accounts. Satellite environmental accounts serve the basic functions of a national accounting system: they provide the raw material needed by policy makers, businesses, and individuals to track important trends and to determine the economic importance of changes in environmental variables. In addition, developing environmental satellite accounts allows experimentation and encourages the testing of a wide variety of approaches.

The panel recommends that the core income and product accounts continue to reflect chiefly market activity. Given the current state of knowledge and the preliminary nature of the data and methodologies involved—especially in areas related to nonmarket activities—developing satellite or supplemental environmental and natural-resource accounts is a prudent and appropriate decision. (Recommendation 5.5)

BEA'S RESUMPTION OF NATURAL-RESOURCE AND ENVIRONMENTAL ACCOUNTING

Given the importance of augmented accounts, the panel is concerned that, particularly since BEA's work in this area stopped in 1994, the United States has fallen behind in developing environmental and other augmented accounting systems. The United States has in place today only the barest outline of a set of natural-resource and environmental accounts, with numerical estimates limited to subsoil mineral assets. This lag has occurred even as the importance of the environment has become increasingly obvious.

The panel recommends that Congress authorize and fund BEA to recommence its work on developing natural-resource and environmental accounts, and that BEA be directed to develop a comprehensive set of market and nonmarket environmental and nonenvironmental accounts. (Recommendation 5.3)

ACCOUNTING FOR SUBSOIL MINERAL ASSETS

The first phase of BEA's work on integrated environmental and economic accounts, published in 1994, resulted in a full and well-documented set of subsoil mineral accounts with useful estimates of the value of mineral reserves. This effort reflects a serious and professional attempt to value subsoil mineral assets and assess their contribution to the U.S. economy. BEA's methods are widely accepted and used by other countries endeavoring to extend their national income accounts, and BEA has relied on sound and objective measures in developing these prototype accounts. BEA should be commended for its initial efforts to value subsoil mineral assets in the United States.

The panel's review uncovered a number of issues that arise in the valuation of mineral resources. The most important issues for further study are the value of mineral resources that are not proven reserves, the impact of ore-reserve heterogeneity on valuation calculations, the distortions introduced by associated capital and production constraints, the volatility in the value of mineral assets introduced by short-run price fluctuations, and the differences between the market and social value of subsoil mineral assets.

From a substantive point of view, the subsoil mineral accounts provide a useful summary of trends in the value of subsoil mineral assets. Two important findings from the initial IEESA are that subsoil assets constitute a relatively small portion of total U.S. wealth and that real proven mineral wealth (resources and associated assets) has remained roughly constant over time. These are important and interesting results that were not well established before BEA developed the subsoil mineral accounts.

While subsoil assets currently account for only a small share of total wealth in the United States, and their depletion does not appear to pose a threat to sustainable economic growth, this situation could change in the future. A good system of accounts could address the widespread concern that the United States is depleting its mineral wealth and shortchanging future generations. By properly monitoring trends in resource values, volumes, and unit prices, the national economic accounts can identify the condition of important natural resources, not only at the national level, but also at the regional and state levels. Better measures would also allow policy makers to determine whether additions to mineral reserves and capital formation in other sectors are offsetting depletion of valuable minerals. Development of reserve prices and unit values would help in assessing trends in resource scarcity. Comprehensive mineral accounts would provide the information needed to develop sound public policies for mineral resources, including minerals on public lands.

Other countries and international organizations are continuing to develop accounts that include subsoil assets and other natural and environmental resources. The United States has historically played a leading role in developing sound accounting techniques, exploring different methodologies, and introducing new approaches. Unfortunately, the United States has lagged behind other countries in developing natural-resource and environmental accounts since BEA's work in this area stopped. Resumption of BEA's work on augmented accounting would allow the United States to exercise leadership in the manner in which such accounts are developed internationally.

Improved mineral accounts at home and abroad would provide substantial economic benefit to the United States. Improved accounts would be particularly useful for those sectors in which international trade is important. Indeed, as is evident from recent cataclysmic events in financial markets—such as the Mexican crisis of 1994-1995 and the financial crises of East Asian countries in 1997-1998—U.S. interests suffer when foreign accounting standards are poor. The United States is a direct beneficiary of better accounting and reporting abroad. Better international mineral accounts would improve understanding of resource consumption and production trends abroad and help in assessing the likelihood of major increases in the prices of oil and other minerals of the kind witnessed in the 1970s. Improved accounts at home and abroad would allow governments and the private sector to better forecast and cope with the important transitions in energy and materials use that are likely to occur in the decades ahead. To the extent that the United States depends heavily on imports of fuels and minerals from other countries, it would benefit from better mineral accounts abroad because the reliability and cost of imports can be more accurately forecast when data from other countries are accurate and well designed.

   The panel recommends that BEA develop and maintain a set of accounts for domestic subsoil mineral assets. (Recommendations 3.9 and 3.10)

ACCOUNTING FOR RENEWABLE AND ENVIRONMENTAL RESOURCES

   BEA had not yet begun developing its accounts for renewable and environmental resources when Congress suspended the agency's work on environmental accounting. Environmental accounting is a useful way to represent interactions between market activity and the environment. There are three major types of interactions: additions and depletions of natural resources that occur when minerals and energy resources are discovered or mined, when timber grows or is harvested, and when groundwater is withdrawn or replenished; alterations in the quality of the natural environment that occur when the composition of air, water, or soil changes; and expenditures made to reduce the effect of economic activities on the environment. The main value of natural-resource and environmental accounting is to illuminate the full role played by these resources in the economy.

   The panel recommends that BEA continue its work on developing accounts for renewable natural resources and the environment. (Recommendation 4.1)

Valuing environmental goods and services requires distinguishing between private goods and public goods. Private goods can be provided separately to different individuals with no external benefits or costs to others; public goods have benefits or costs that are spread indivisibly among the entire community or even the entire planet. Price data are relatively reliable for private market goods, such as the timber produced from forestry assets. Values for near-market goods—such as freely collected firewood—can be constructed by comparing the near-market goods with their market counterparts. Techniques for valuation of public goods are still under development. Some techniques—such as hedonic-price or travel-cost studies—rely on behavioral or market-based estimates; while these estimates are subject to significant measurement problems, they are conceptually appropriate in economic accounts. Other techniques, such as contingent valuation, are not based on actual behavior, are highly controversial, and are subject to potential measurement errors. An important issue here (as it is throughout the federal statistical system) is developing measures of accuracy, both for satellite accounts and the main accounts.

For valuation, BEA should rely whenever possible on market and behavioral data. However, novel valuation techniques will be necessary for the development of a comprehensive set of nonmarket accounts. (Recommendation 5.7)

Quantitative data on many market and near-market activities are at present comparatively adequate. Quantitative data for natural resources are often reliable because in many cases there are well-established conventions for their measurement. Quantitative data on some near-market activities, such as the collection of fuel wood for private use and recreational fishing, are conceptually straightforward, and many of these data are currently collected by federal agencies. Quantitative data on other marketable goods, such as stocks of commercial fish, could be improved substantially. The measurement of quantities for nonmarket goods and services, particularly those that have public-good characteristics, suffers from severe methodological difficulties and insufficient data. There are relatively good physical data on emissions of many residuals from industrial and human activities, but there is very little systematic monitoring of human exposures to most harmful pollutants. The data on many environmental variables are currently poorly designed for the construction of environmental accounts.

The panel recommends a concerted federal effort to identify and collect the data needed to measure changes in the quantity and quality of natural-resource and environmental assets and

associated nonmarket service flows. Greater emphasis should be placed on measuring effects as directly as possible, particularly on measuring actual human exposures to air and water pollutants. (Recommendations 4.3 and 5.9)

True public goods, such as biodiversity, species preservation, and national parks, present major conceptual difficulties for incorporation into a national accounting system. More work will be needed on techniques for measuring production flows and values for the assets and services of true public goods in order to make them compatible with the prices and quantities used in the core accounts.

Notwithstanding the awesome difficulties that arise in accounting for air quality, this is likely to be the single most significant sector in environmental accounts. Creating accounts for sectors such as clean air is an essential component of efforts to develop a comprehensive set of nonmarket accounts. However, the construction of air-quality accounts transcends the present scope and budget of BEA and will require further research on the underlying physical phenomena, measurement methods, and economics.

BUDGETARY IMPLICATIONS

The cost to BEA and other agencies of developing and maintaining a set of augmented accounts will depend on the intensity and extent of the effort. The costs would be small for a minimal program of incremental improvements limited to a few natural-resource sectors. Estimates provided by BEA indicate that the cost of a small activity, including reinstatement of the pollution abatement survey, would be approximately $1.5 million annually. Developing a comprehensive set of environmental and augmented accounts would require more funds over a longer period. Although the cost of a comprehensive accounting system will depend on the extent to which BEA is able to draw on data and expertise from other agencies, a preliminary estimate is that a full set of accounts would require incremental outlays for BEA and other agencies of about $10 million per year for a decade or more.

SUMMARY

In weighing future directions for environmental and augmented accounting in the United States, the panel concludes that developing a comprehensive set of nonmarket accounts is the most promising approach. Because of the high cost and low return involved, reliable nonmarket accounts will not be supplied by the private sector. In a country as large, complex, and wealthy as the United States, providing information on the structure and interactions of the economy and the environment is an essential function of government, and one the federal government is supporting insufficiently at present.

Developing a comprehensive set of nonmarket accounts for the United States is a large undertaking that would stretch the scope and specialized expertise of BEA. Moreover, if undertaken with the currently projected available resources, such a task would clearly result in cutting back other important BEA functions and proposed improvements. The panel is mindful of BEA's important mission and of the invaluable nature of the data it provides on marketed economic activity. In addition to furnishing key macroeconomic data and information on different sectors of the economy, BEA has been highly innovative in introducing new data and approaches.

The panel concludes that the development of environmental and natural-resource accounts is an essential investment for the nation. It would be even more valuable to develop a comprehensive set of environmental and nonmarket accounts. The panel emphasizes, however, that environmental accounts must not come at the expense of maintaining and improving the current core national accounts, which are a precious national asset. (Recommendation 5.8)

1

Introduction

The last quarter-century has seen increasing awareness of the interactions between human societies and the natural environment in which they thrive and upon which they depend. This awareness has been heightened by concerns about resource scarcity, environmental degradation, and global environmental issues. The combination of increased awareness of the environment and recognition of the primitive state of much of the nation's environmental data has led to a widespread desire to supplement U.S. national economic accounts to include natural resources and environmental assets. The idea of including environmental assets and services in the national economic accounts is part of a larger movement to develop broader economic indicators. This movement reflects the reality that economic and social welfare does not stop at the market's border, but extends to many "near-market" and nonmarket activities, such as household production, leisure activities, and environmental quality.[1]

---

[1]"Near market," "natural resources," "environmental assets," and other major terms used in environmental accounting are defined in the glossary (Appendix D).

THE NATIONAL INCOME AND PRODUCT ACCOUNTS

Concepts

The modern national income and product accounts are among the great inventions of the twentieth century. Among other things, they are used to judge economic performance over time, to compare the economies of different nations, to measure a nation's saving and investment, and to track the business cycle. Much as satellites in space can show the weather across an entire continent, the national accounts can give an overall picture of the state of the economy.

This report addresses the question of whether the U.S. economic accounts should be extended to include activities involving natural resources and the environment. It will be useful at the outset to explain what is meant by "accounting" and by the "national income and product accounts." In its most general sense, the purpose of accounting is to provide economic information about a household, organization, or government. Accounts are generally divided into "income accounts," which record receipts and outlays during a given period such as a year, and "asset accounts," which provide a snapshot of the assets, liabilities, and net worth of an entity at a given date. People are most familiar with the income accounts and balance sheets of businesses, but the same concepts apply equally well to individuals, governments, and nations.

The present report is concerned with a specific set of accounts known as the National Income and Product Accounts (NIPA). The fundamental purpose of the NIPA is to provide a coherent and comprehensive picture of the nation's economy. These accounts measure the total income and output of the entire nation, including households, business and not-for-profit enterprises, and different levels of government. The key elements of the NIPA—what this report calls the "core accounts"—measure the total market output and income of the United States. The most important item is gross domestic product (GDP), a measure of the nation's total output of goods and services and the total income of the nation generated by that output. GDP represents the sum of the dollar values of consumption, gross investment, government purchases of goods and services, and net exports produced within the nation during a given year. It also represents the income earned as wages, profits, and interest, as well as indirect taxes. In addition to the totals for the nation, the NIPA provide a rich array of data on output and incomes in different industries and regions, as well as a record of international transactions.

To date, the major focus of the U.S. national accounts has been on developing income accounts, with relatively less attention on asset accounts. In addition, a central organizing principle of the accounts is that national output is, with a few exceptions, defined to be the production that is reflected in the sales and purchases of the market economy. Indeed, the NIPA's omission of many nonmarket activities—particularly those involving natural resources and the environment—along with the potential distortion in measures of national output and wealth stemming from that omission, is the very concern that led to the environmental accounting addressed in this report.[2]

History

National accounts were first developed by Sir William Petty in 1665, with estimates being primarily the work of individual scholars until World War I.[3] There was little appreciation during this period of economic statistics as a public good. Moreover, although there were sporadic federal efforts to develop estimates of national income and output, the impetus for systematic development of the accounts came during the Great Depression. Measures of national output at that time were incomplete and produced with a considerable lag, so policy makers had only an impression of economic trends. The lack of reliable and timely data led to a congressional resolution during the Great Depression, introduced by Senator Robert La Follette:

> RESOLVED, That the Secretary of Commerce is requested to report . . . estimates of the total national income of the United States for each of the calendar years 1929, 1930, and 1931, including estimates of the portions of national income originating from [different sectors] and estimates of the distribution of the national income in the form of wages, rents, royalties, dividends, profits, and other types of payments.

The first set of accounts was developed at the Commerce Department under the leadership of Dr. Simon Kuznets, who received the Nobel prize for his pioneering role in that work. The effort was conducted in collaboration with the National Bureau of Economic Research, a private nonprofit economic research organization. The resulting set of accounts was submitted to the Senate in 1934 and published as a Senate document.

The major aggregates of the national accounts—including gross national product (GNP) and the division between consumption and invest-

---

[2]For a description of the methodology underlying the U.S. NIPA, see Bureau of Economic Analysis (1995b).

[3]The historical discussion that follows is based on Carson (1975).

ment—date from Kuznets's work in the 1930s. The NIPA aggregates are analogous to a firm's income statement in that they represent economic activity for a period of time, usually a quarter or a year. Over the next decade the accounts were elaborated and redefined. The basic framework delineated in 1947 is discussed in Kuznets (1948c) and has, with a few exceptions, remained virtually intact since that time. The major aggregates today are GDP and expenditure, national income, personal income, and personal disposable income. The nation's asset accounts, analogous to a firm's balance sheet, have also been developed as part of the national accounts. The most developed is a set of capital asset accounts, reflecting a component of the nation's wealth.

AUGMENTED NATIONAL ACCOUNTS

Background

As noted earlier, the traditional national accounts include primarily the final output of marketed goods and services—that is, of goods and services that are bought and sold in market transactions. Notwithstanding the importance of the traditional accounts, it has long been recognized that limiting them to market transactions distorts them as a measure of economic activity and well-being. A vast and rapidly changing amount of nonmarket activity produces goods and services that are quite similar to those produced in the marketplace, but are omitted from traditional accounts. Time spent cooking hamburgers at Wendy's is counted in the national accounts, while cooking time at home is not; nannies' services are reckoned as part of GDP, while mommies' and daddies' services are not; the value of swimming in a commercial swimming pool is captured by GDP, while the value of swimming in a public lake or in the ocean is not.

In response to growing concerns about the accuracy of traditional measures of economic activity, many efforts have attempted to broaden the traditional accounts to include important sectors of nonmarket activity beyond the imputations of rent on owner-occupied housing, certain financial services, and the value of home-grown food, all of which were in the earlier accounts. The history of augmented accounting, some of which includes adjustments for the environment, goes back to the early 1970s (Eisner, 1971). Most of the early efforts were undertaken by private scholars. Significant examples of sectors examined in studies addressing extension of the accounts include household production and unpaid work, the services of consumer durables, research-and-development capital, leisure time, and informal and home education. In most countries, however, few efforts were made to broaden the official national accounts until the 1980s.

Although many different approaches have been taken, the guiding principle in augmented economic accounts is to measure as much of economic activity as is feasible, regardless of whether it takes place inside or outside the marketplace. Augmented national economic accounts are designed to provide better measures of final output—including what consumers currently enjoy in the way of goods and services, as well as the accumulation of capital, of all kinds, that will permit the future production of goods and services.

A set of well-designed augmented accounts can overcome the recognized shortcomings of the current market-based accounts. Environmental accounts can provide information useful for managing the nation's public and private assets, for improving regulatory decisions, and for informing private-sector decisions. Data on comprehensive income and output are a public good that would benefit the nation even though individual firms might not profit from building such accounts. The collection of these data is an investment that would have a high economic return for the nation because better information would allow both the public and private sectors to make better decisions. There are many examples of how comprehensive economic accounts can bring economic benefits. These include better estimates of the impact of regulatory programs on productivity, improved analyses of the costs and benefits of environmental regulations, and more effective management of the nation's public lands and resources.

Augmented national accounts would also be valuable as indicators of whether economic activity is sustainable. From the point of view of a national economy, sustainable national income is usefully defined as the maximum amount a nation can consume while ensuring that all future generations can have living standards at least as high as those of the current generation. The NIPA have a close relationship with measures of sustainable income. The usual measure of net domestic product (NDP) corresponds to the highest sustainable level of consumption under certain special conditions. The most important of these conditions are the inclusion of all segments of consumption and net investment—whether market or nonmarket—and the absence of technological change or other dynamic autonomous elements.

It is clear that the national productivity depends on many nonmarket elements, including not only the environment, but also such things as schooling, health care, and social capital in volunteer and civic organizations. It may not be possible to capture all these important facets of modern society in the nation's accounts, but an attempt should surely be made to include those which are clearly related to economic life, can be measured with sufficient precision, and present a more accurate picture of the nation's economic activity.

Integrated Environmental and Economic Satellite Accounts (IEESA) and the Congressional Mandate

The U.S. Bureau of Economic Analysis (BEA) has studied augmented accounting since the early 1980s. BEA began work on the U.S. version of environmental accounting, known as Integrated Environmental and Economic Satellite Accounts (IEESA), in 1992. This work was given additional impetus when President Clinton put environmental accounting on a fast track in his 1993 Earth Day speech by stating: "Green GDP measures would incorporate changes in the natural environment into the calculations of national income and wealth."

BEA produced its first set of IEESA, along with a proposed framework for further developing the accounts, in 1994 (see Bureau of Economic Analysis, 1994a). A three-phase work plan was proposed. The first phase, with preliminary results presented in the April 1994 *Survey of Current Business*, involved delineating the overall framework and developing a set of prototype satellite accounts for subsoil assets such as oil, gas, and major nonfuel minerals. The second phase would extend the accounts to renewable and other natural resources such as trees on timberland, fish stocks, and water resources. The third phase would involve nonmarket environmental assets, including the economic value of the degradation of clean air and water and the value of recreational assets such as lakes and national forests. (For a discussion of the work plan and the preliminary results, see Bureau of Economic Analysis 1994a, 1994b.)

Congressional concerns about environmental accounting were raised shortly after the initial publication of the draft IEESA. As a result, in the committee report accompanying appropriations for the Department of Commerce in fiscal year 1995, Congress directed that the department suspend further work on the IEESA until the methodological issues involved had been reviewed:

> The Committee is concerned about the Administration's initiative on "Green GDP" or "Integrated Environmental-Economic Accounting," which seeks to provide a measurement of the contribution of natural resources to the Nation's economy. The Committee recognizes that there may be value to the measurement proposed to be taken under this initiative, but has concerns as to whether the Department has adequately addressed the questions of appropriate methodology and proposed applications of the data in developing this initiative. The Committee ex-

pects the Department to suspend its work on this initiative until a more thorough analysis of the proposed methodology and applications of Green GDP can be undertaken by an independent entity. (House Report Accompanying HR4603, FY 1995, for the Department of Commerce)

CHARGE TO THE PANEL

In response to the above congressional mandate, the Commerce Department asked the National Academy of Sciences to undertake a review of environmental accounting. The Panel on Integrated Environmental and Economic Accounting, working under the aegis of the Committee on National Statistics, was established to perform this review. The Academy's charge to the panel was as follows:

A panel is planned to examine the objectivity, methodology, and application of integrated environmental and economic accounting in the context of broadening the national economic accounts. The panel would review the approaches by BEA and others to the valuation of environmental resources, recommend improvements, and suggest further research that would strengthen the knowledge base about valuation. A panel of about 12 members would be convened of specialists in national income accounting, in particular in some areas covered by the augmented accounts, such as private sector accounting, natural resource economists, and relevant environmental scientists. The panel would meet about five times over a two-year period. It would conduct three major reviews:

1. The panel would review the proposed revisions in general to broaden the national accounts and examine progress made by other national statistical agencies to introduce augmented accounts in the environmental and other areas. The panel would review international efforts on the valuation of environmental resources, particularly in Canada and Western Europe, and review theoretical and empirical work by private agencies and scholars.

2. The panel would review the first phase of BEA's augmented environmental accounts, which primarily include revisions of the accounts to incorporate reduction of subsoil assets, such as oil and gas.

3. The panel would review plans and methodology proposed by BEA for its second phase on renewable appropriable resources, such as water and timber, and for its third phase on environmental resources, such as clean air.

The panel would compare the methodologies with research in other countries and in nongovernmental research, advise BEA on some of the strengths and weaknesses of different approaches, and recommend improvements and needed research.

This report details the panel's findings and recommendations. The central issues examined are whether BEA's IEESA are useful for the United States and whether work on the IEESA should resume.

ORGANIZATION OF THIS REPORT

Chapter 2 considers the importance of integrating environmental accounts with the NIPA, and reviews alternative approaches to such accounting. The discussion includes a detailed examination of the theoretical rationale behind extending the NIPA to include all market and nonmarket economic activity.

Chapter 3 details the extension of the accounts to subsoil mineral assets, such as fossil fuels and minerals. This was the first area (beyond accounting for pollution abatement capital expenditures) in which BEA addressed environmental matters; it is an area about which Congress has expressed concern; and it provides an excellent introduction to the questions and problems associated with the IEESA.

The extension of the IEESA to renewable and other natural resources, as proposed by BEA for its Phase II effort, is covered in Chapter 4. After examining BEA's work in this area, the chapter offers two extended examples—forests and clean air—to illustrate opportunities and problems that arise in developing such accounts.

Finally, Chapter 5 presents the panel's overall appraisal of environmental accounting in the United States, as well as the panel's conclusions and recommendations.

2

The National Income and Product Accounts: History and Application to the Environment

NATURE'S NUMBERS

Natural and social scientists concerned about natural resources and the environment have endeavored to take the measure of nature. Measures used for this purpose range from those used to monitor the state of major environmental indicators, such as air and water quality, to analytical measures of major environmental aggregates. For the most part, however, measures of the economic contribution of natural resources and the environment have lagged behind physical measures. The slow development of economic measures is due to two major factors. First, economic accounts generally record and measure activities that pass through the marketplace, while most of the activities that raise environmental concerns—from air pollution to appreciation of pristine wildernesses—take place outside the market. Second, the paucity of data and difficulties of valuation for most environmentally related activities make constructing economic measures much more difficult than is the case for market-related activities. The end result is that most nations produce detailed national economic accounts accompanied by vast quantities of useful data for market-related activities and little or no comparable data for nonmarket environmental activities.

The intuitive idea behind the desire to broaden the U.S. national accounts is straightforward. Natural resources such as petroleum, minerals, clean water, and fertile soils are assets of the economy in much the same way as are computers, homes, and trucks. An important part of the economic picture is therefore missing if natural assets are omitted in creating the national balance sheet. Likewise, consuming stocks of valuable subsoil assets such as fossil fuels or water or cutting first-growth forests is just as much a drawdown on the national wealth as is consuming aboveground stocks of wheat, cutting commercially managed forests, or driving a truck.

HISTORY OF AUGMENTED ACCOUNTING

General Developments

From the perspective of environmental accounting, the key point to recognize is that gross domestic product (GDP) is conceptually defined to include only the final output of marketed goods and services, that is, goods and services that are bought and sold in market transactions. This point is clearly stated in a comprehensive discussion of the National Income and Product Accounts (NIPA): ". . . the basic criterion used for distinguishing an activity as economic production is whether it is reflected in the sales and purchase transactions of the market economy" (U.S. Department of Commerce, 1954).

There are, however, important exceptions to this basing of the NIPA on market transactions. One is the exclusion of illegal activities such as drugs, prostitution, and illegal gambling; thus GDP will rise as gambling moves into the legal market sector. In addition, there are imputations for near-market services that are not recorded in market transactions. For example, there is an imputation for the services of owner-occupied housing so that these services can be included in output and income as are the rent and output associated with rental housing. There is also an imputation for the fuel and food produced on farms and consumed by the farmers themselves. A further large imputation is made for banking and other financial services furnished by financial businesses below cost in lieu of interest payments.

The key issue involved in environmental and other augmented accounts is whether to broaden the above boundaries and if so, how and how far. It has long been recognized that drawing the line at the limits of the market distorts the value of the NIPA as a measure of economic activities and well-being (see also Chapter 1). There is a vast and changing amount of productive nonmarket activity that produces goods and services quite similar to those produced in the marketplace. Commercial laundry services are reckoned as part of GDP, while parents' laundry services are not; the value of downhill skiing at a ski area is captured by GDP, while the value of cross-country skiing in a national park is not.

At the same time, while recognizing the importance of considering alternative measures, it is essential to retain the conventional market-based accounts as a central component of our national accounts. These *core accounts* are of great importance for purposes of historical and international comparison and will continue to be a critical indicator for economic policy making. The objective of augmented accounting is not to replace the core accounts with a preferred new bottom line; rather, the emphasis is on developing alternative approaches and measures that can illuminate the diverse dimensions of economic activity.

Work on augmented accounting by official statistical agencies, as well as by individual scholars, has provided estimates for a wide variety of nonmarket activities for experimental augmented national accounts (see Eisner, 1988, for a comprehensive review of augmented accounting). Beyond the environmental arena, which is reviewed in the next section, significant examples of work to extend the accounts include the following areas:

- The value of home production and unpaid work
- The value of the services of consumer durables (similar to the imputation of services of owner-occupied housing)
- The value of research-and-development capital
- The value of leisure time
- The value of informal and home education

This work on extending the accounts is motivated by the idea that expanding the boundaries of the accounts would provide a better estimate of the size, distribution, and growth of economic activity and economic welfare than that offered by the current accounts.[1]

In revising and extending the U.S. NIPA, the Bureau of Economic Analysis (BEA) is following guidelines suggested by the internationally formulated and recommended U.N. System of National Accounts (SNA) (Parker, 1996 and 1991). These efforts have entailed both modifications in core measures, such as the introduction of separate current and capital accounts for government, and the development of satellite accounts, such as for research and development. *Satellite accounts* (sometimes referred to as supplemental accounts) expand the analytical capacity of the national accounts without overburdening them or interfering with their general orientation. Because they supplement rather than replace the core accounts, they can serve as a laboratory for experimentation and provide a means for applying alternative approaches and new methodologies.

---

[1]Many of these examples are reviewed in Eisner (1988).

The guiding principle in developing augmented accounts is to measure as much economic activity as is feasible, regardless of whether that activity is of a market or nonmarket nature. The goal is to achieve a better measure of final output—of what consumers in the United States currently enjoy in the way of goods and services, and of the accumulation of capital, of all kinds, that will permit the future production of goods and services.

In terms of current consumption, augmented output includes not merely what consumers buy in stores, but also what they produce for themselves at home; the government services they "buy" with their taxes; and the flow of services that are produced by environmental capital such as forests, national parks, and ocean fisheries. The need to include nonmarket components arises because of the trade-offs between market and nonmarket activity. For example, parents produce more in the market when they go to work, but they also have less time at home for child care and domestic services. Likewise, the resources used to provide government services that add to real consumption may reduce the quantity of services provided by businesses. In the environmental area, resources devoted to removing lead from gasoline and paint will lower conventionally measured consumption, but will raise the nation's human capital by protecting children from brain damage and other debilitating illnesses.

Similar issues arise in the measurement of national saving and investment. Conventional NIPA saving and investment measures include only tangible investments in plant, equipment, and inventories. This conventional picture omits the much larger intangible and human investments in education, training, research and development, health, and the environment. A complete set of accounts would entail full integration of comprehensive investment flows with comprehensive capital or wealth accounts. These accounts would then relate not only to current production of goods, but also to changes in the value of human capital; to the accumulation of knowledge and technical advances; and to the improvement or deterioration of the basic environmental capital of land, water, and air. Development of a complete set of capital accounts would thus give the nation a much more complete picture of how well the current generation is performing in its role as trustees of the nation's tangible, human, and natural resources.

Comprehensive accounts and environmental accounting provide information that can help governments set sound economic and social policies and aid the private sector in making productive investments. An important example is use of pollution abatement costs to estimate the impacts of regulation on productivity and output growth. Studies by Denison (1979) and by Jorgenson and Wilcoxen (1990) have provided valuable information on the relative importance of regulation, pollution control expenditures, and other factors in the slowdown of productivity growth in the United States after 1973. In addition, these data have been crucial inputs to studies of the cost of air pollution regulation and the benefits and costs of controlling air pollution conducted by the U.S. Environmental Protection Agency.

Two major issues arise in the design of augmented accounts. The first is where to draw the line when extending the accounts beyond the boundary of market transactions. The dilemma is similar to that faced by the little boy who said, "I know how to spell banana, but I don't know where to stop." Should the line be drawn at near-market activities—for example, home-cooked hamburgers, depletion of oil and timber resources, fish caught and consumed by anglers, and services of consumer capital such as automobiles and washing machines? Or should the accounts extend to all private goods, such as educational investments and the value of visits to Yellowstone National Park? Should the accounts attempt to measure the value of leisure time? Should they extend to include public goods such as the value of clean air and clear water? Should they include international concerns such as the damages from ozone depletion and global warming? These thorny questions are taken up later in this report, but we note here that they are pervasive in the design of augmented accounts.

The second major issue is how to measure nonmarket activities. Measurement involves collecting data that will support estimates of both quantities and prices. While data on market activities are often costly to collect, for the most part the elemental data exist in the form of individual transactions in which someone buys a banana, a computer, or a haircut—transactions that are generally recorded. Nonmarket activities pose difficulties because the physical activities involved are generally not recorded, and there are no objective records of the valuations. An example of the difficulty is a consumption service such as swimming in the Atlantic Ocean. No one records how many times Americans actually swim in the Atlantic Ocean in a given year. More difficult is the valuation of swimming: since swimming is generally free, except in congested areas, we do not know how to value the swims. There are numerous techniques available for estimating both the quantity and value of nonmarket activities such as swimming, but they almost always require gathering additional data and involve complex imputations of value where no market data are available.

We must not, however, forsake what is relevant and important merely because it presents new problems and difficulties. The economic light is brightest under the lamppost of the market, but neither drunks nor statisticians should confine their search there. In extending the accounts, we must endeavor to find dimly lit information outside our old boundaries of search, particularly when the activities are of great value to the nation.

Developments at the Bureau of Economic Analysis

Over the last decade, BEA has taken a number of important steps in extending the core economic accounts and developing satellite and supplemental accounts (BEA, 1995a). Among the most important developments in the core accounts are the following:

• *Improved measures of price and output.* BEA has pioneered the use of improved measures of price and output, including the use of chain-weighted price and output indexes. These measures allow more accurate tracking of inflation and output than did earlier fixed-weight measures. This work has demonstrated that these state-of-the-art concepts can be implemented routinely in national statistical measures.

• *Improved investment accounts.* BEA has moved to broaden the U.S. investment accounts in line with international standards by introducing estimates of government investment and capital and improving the estimates of depreciation and capital stocks.

• *Improved international accounts.* Recognizing the growing importance of services in the nation's economy, BEA has incorporated new information on international trade in services and revised estimates of foreign direct investment.

In its efforts to improve the national economic accounts, BEA has been proceeding in a prudent and conservative fashion, employing proven and consistent techniques. In its core national accounts, BEA employs the concept of Hicksian income (see Appendix A). Such production-based measures of income and output are useful for delineating market activity and should continue to form the basis of the core national accounts.

Market-based concepts are inadequate, however, for tracking the entire range of economic activity, market and nonmarket. The purposes of augmented accounting are to provide more comprehensive measures of output, saving, and investment; to ensure that the accounts treat economic activity in a consistent way when the boundaries between market and nonmarket activities change; and to provide information on the interaction between the economy and the environment so that natural and environmental resources can be more effectively managed and regulated.

IMPORTANCE OF ENVIRONMENTAL AND NATURAL-RESOURCE ACCOUNTING

Environmental and natural-resource accounting has emerged over the last three decades in response to increasing awareness of the interaction between the natural environment and economic activity. Growing concerns about resource scarcity were reinforced by the dramatic increases in energy and mineral prices of the 1970s. Many began to worry that the nation was rapidly depleting its precious stocks of subsoil assets. Further awareness resulted from documentation of the economic and social costs of environmental degradation and pollution in terms of human health and property values, reinforced by pictures of rivers and lakes on fire and serious oil spills.

A set of well-designed environmental accounts could overcome the shortcomings of the current market-based accounts. Indeed, the construction of environmental accounts is one element of the more general task of developing a set of comprehensive economic accounts that includes both market and nonmarket economic activity. This section reviews the primary shortcomings of the current accounts and explains why a well-constructed set of comprehensive accounts would have significant economic value to the nation.

Deficiencies of Current National Accounts

Efforts to develop alternative accounting approaches to supplement the standard market accounts with measures of changes in consumption and investment in natural resources and the environment have been undertaken in response to three perceived deficiencies in the way the conventional accounts treat natural resources and the environment.

First, as an indicator of economic well-being, the accounts sometimes behave perversely with respect to environmental degradation and changing stocks of natural resources. For example, cutting down the nation's dwindling redwood forests increases GDP, yet no account is taken of the loss of this precious asset because the nation's forests have not been included in the asset accounts. For similar reasons, when fishing activities increase the harvest of cod or halibut, the national accounts record the increased production and consumption, but omit the decline in breeding stocks and the costs imposed on future producers and consumers. And pollution abatement expenditures increase measured output—even when such expenditures serve only to offset environmental deterioration, and there is no net increase in current or future consumption. In these and many other examples, changes in production do not reflect genuine changes in economic well-being and may even result in economic harm or cost in the future.

Second, the standard national accounts are inconsistent in their treatment of different forms of wealth. For example, the NIPA include a full set of accounts of gross investment, net investment, depreciation, and the capital stock for produced, tangible producer capital. In contrast, natural capital—such as oil and gas deposits, forests, soils, and underground aquifers—is largely omitted from the accounts. When a commercially grown tree is cut, the production cost of the tree is counted as a cost of production, but when a first-growth national forest is clear-cut, there is no parallel subtraction. As a factory ages, this is counted as a depreciation charge, but the accounts are not charged when an oil deposit is exhausted. Likewise, the national accounts nowhere reflect the occurrence of widespread deterioration or improvement in the quality of environmental assets such as air and water. The distinction between gross and net investment for reproducible capital is justified on the grounds that those investments which simply replace depreciated stock add nothing to economic well-being and that failure to subtract depreciation would yield income measures that might be unsustainable in the long run. The logic of this argument is equally applicable to environmental and natural capital.

The third and perhaps most important deficiency of the conventional national accounts is that they give a very incomplete picture of the full scope of economic activity. By focusing only on marketed outputs and factors of production, the conventional accounts neglect a large number of economically significant inputs and outputs that are not bought and sold in markets. In the environmental area, these nonmarketed inputs and outputs often include the free goods and services provided by environmental assets such as air, water, forests, and complex ecosystems. Many of these assets—such as recreational sites in Yellowstone Park, stocks of underground water and flows of river water in the Southwest, and public parks in New York City—are limited or fixed in supply. Thus they have economic scarcity value even though they may lack market prices. Because the conventional accounts omit such economically valuable but nonmarketed goods and services, they overstate the role of market inputs and outputs in economic welfare. They also fail to provide business, citizens, and policy makers with the full and accurate assessment of the state of economic activity that is needed for economic policy and rational environmental management.

Finally, it should be emphasized that the current NIPA do not focus on a conceptually appropriate definition of market national income and output. The most appropriate measure of national output in the core accounts today is real net national product, which measures the total net output and income accruing to residents of the United States, corrected for inflation. This differs from the measure currently emphasized, real GDP, in two ways. First, GDP includes depreciation or capital consumption, which exaggerates sustainable income by including in national income a sum that cannot sustainably be consumed. Traditionally, output measures have emphasized gross rather than net product because depreciation is difficult to measure accurately. Second, GDP excludes the net factor earnings abroad of domestic residents, which is included in national product. Inclusion of net factor earnings abroad is desirable if output is designed to measure the sustainable consumption of the nation. The recent switch in emphasis from national to domestic product occurred because domestic product is more closely related to domestic output and employment. While the emphasis on GDP rather than net national product is understandable, the panel emphasizes that the latter is conceptually preferable as a measure of sustainable income.

Value of a Comprehensive Set of Accounts: Scorekeeping and Management

Economic accounting—whether it be business accounting or the accounting of a nation's economic activity—traditionally serves two major functions: it offers a way to track the economic performance of a business or a nation, and it provides an organized body of economic data that enhances the ability of an organization or a nation to manage its economic affairs. The principal reason for growing interest in natural-resource and environmental accounting is the belief that improved accounting for the contribution of natural capital will enhance the ability of the conventional accounts to serve both of these functions.

The NIPA are the major way nations keep score on overall, regional, or sectoral economic performance, past and present. The core accounts include production measures such as gross national product (GNP) and GDP, along with data measuring market incomes and a broad array of sectoral data. These core accounts are widely used to gauge a nation's economic performance over time and to compare economic performance among nations; they are an essential tool for assessing the state of the economy and formulating macroeconomic stabilization policy. For example, economic research has shown a close link between movements in GDP and changes in the unemployment rate, changes in tax revenues, and the federal budget deficit. Understanding the economy requires comparing current trends and movements in national output with those of various historical periods in order to forecast the future. This scorekeeping function of the national accounts is widely accepted in spite of many deficiencies in the measures of prices and outputs and the numerous interpretative problems introduced because the core accounts are limited to market transactions (see Hicks, 1940; see also Kuznets, 1948a, 1948b).

As discussed above, measures of augmented national income and product endeavor to extend the purview of the economic accounts by including a broader set of consumption, investment, and income measures. These augmented accounts give a more balanced view of the trends of overall economic activity and provide more accurate estimates of trends in income, saving, and investment. More comprehensive systems that account for negative outputs such as pollution and positive outputs such as outdoor recreation will yield more meaningful indicators of economic performance. One valuable contribution of well-designed comprehensive measures is that they can eliminate anomalies in the national accounts. For example, according to conventional measures of economic performance, oil spills and earthquakes often raise GDP and appear to make the nation better off. Such anomalies would be redressed by appropriate accounting measures. Similarly, improved measures would correct the anomaly that a nation with abundant natural parks and recreational opportunities provided freely to its citizens appears to be worse off than a nation that provides recreation only through commercial theme parks.

A final set of scorekeeping measures relates to what is called "sustainable income." These measures address the question of whether the nation is setting aside sufficient tangible and intangible capital and new technological knowledge to ensure that future generations will have an adequate standard of living. Sustainable national income is defined as the maximum amount a nation can consume while ensuring that future generations will have living standards at least as high as those of the current generation. It turns out that ideal measures of sustainable income are closely related to current national income and product measures. Techniques for measuring sustainable income are discussed later in this section and in Appendix A.

The second function of economic accounting—providing data needed to manage economic activities—requires gathering a systematic record of all the inputs and outputs that characterize an economic system. The management function of accounts and budgets is widely used by both businesses and governments. While scorekeeping indices may tell a business whether it is profitable, details of accounts and budgets are necessary to help the business make better decisions and improve its profitability. Similarly, while government budgets are valuable summary indicators of the overall importance of the government sector in the economy and of government's net contribution to national saving, the most important function of the budgetary accounts is to help Congress and the executive branch direct the day-to-day operations of the federal government and the allocation of federal resources.

The detailed information embodied in the national economic accounts serves a similar function in the management of national economic policy. Even in countries such as the United States that have strong laissez-faire traditions, the economic accounts are an essential input to major economic policy and forecasting models that influence fiscal and monetary policy. At a more detailed level, the data help businesses track their own sectors and forecast their sales and profits, and are useful for a wide variety of economic activities.

Unfortunately, the conventional economic accounts are sometimes deficient for management purposes because of their omission of those inputs and outputs that are not traded in the marketplace. Resource and environmental accounting expands the list of inputs and outputs so that policy makers are in a much better position to develop and analyze policies, especially those that involve interactions between the natural environment and the market economy. Economic accounts expanded to include resource and environmental activity are especially useful for the analysis of major environmental policies and programs that may affect large segments of the economy, such as those related to water allocation or global warming, or for the analysis of nonenvironmental programs that may have substantial environmental consequences, such as interstate highway programs. Without a comprehensive environmental and non–market accounting framework, each policy analysis requires data collection *de novo*. As a result, analyses of environmental programs today are extremely expensive and inefficient. A system of resource and environmental accounts linked with the conventional economic accounts can provide the inputs for a wide variety of policy analyses at relatively low incremental cost.

The remainder of this section describes in detail the primary benefits of a comprehensive set of accounts.

Comprehensive Accounts Give a Complete Picture of Economic Activity

At the most general level, comprehensive economic accounts provide a complete reckoning of economic activity, whether it takes place inside or outside the boundary of the marketplace. As suggested above, economic decision makers need to understand more than the conditions of the marketplace if they are to make sound decisions. Businesses clearly need and want to know about basic economic conditions in the world, the nation, their region, and their industry. Without such information, firms are flying blind. They run the risk of continuing unsustainable programs to the point of serious decline or even bankruptcy. States and localities similarly require comprehensive accounts of economic activity. Such comprehensive accounts need to include natural-resource and environmental accounting. A firm will pause, for example, before building a plant for which a fuel that is running into short supply would be required or locating in a region whose water supplies are severely limited. Companies may want to build in areas that have many amenities and high environmental quality—conditions that result both from market investments by the public and private sectors and from well-maintained natural capital. Managers and stockholders want to know the unpriced environmental costs of their actions because society may eventually make them pay those costs. These kinds of questions are equally vital for states, localities, and foreign investors, as well as for individuals who face personal investment or locational decisions. In short, environmental accounts are an important tool for providing the information necessary to track economic conditions and make sound decisions.

Limiting the national accounts to market sectors can produce misleading information on overall economic trends. One important example is standard measures of national saving and investment, which include only investment in tangible capital such as factories, equipment, inventories, and houses. By omitting market and nonmarket investments in intangible and human capital, the current national economic accounts can underestimate national saving by almost 500 percent.[2] Another example of misleading signals is in the treatment of the movement of people from unpaid to paid work. Because the unpaid work is not counted in national product while the paid work is, measured national output rises more than the actual amount of total national output of goods and services as labor force participation rises.

Similar issues arise within the natural-resource and environmental sectors. When companies discover large deposits of oil, gold, and other mineral assets, these deposits are not counted among the nation's investments or as an increase in its stock of assets. Similarly, although forests contribute greatly to the nation's well-being, only the timber value of forests is counted as part of the official national output. The value of hunting, fishing, and other forms of nonmarket forest recreation likewise is not counted as part of the national output, even though the total economic contribution of these nonmarket outputs probably exceeds the value of timber production.

The largest distortion in the environmental area probably arises in those sectors related to environmental quality. Economic studies reviewed in Chapter 4 of this report indicate that the nation is devoting more than $100 billion annually to pollution abatement and control expenditures. Yet virtually all the economic benefits from these expenditures are omitted from the national accounts. Even though investments in

---

[2]This underestimation is due to expanding the measures of investment to include acquisition of tangible nonhuman capital by households, acquisition or development of land, expenditures for research and development, expenditures for education, opportunity costs of students' time, expenditures for health, and revaluations of existing assets and liabilities.

clear air and water produce benefits in improved health of the population, improved functioning of ecosystems, enhanced recreational opportunities, and lower property damages, a large share of these benefits is not likely to be captured by the current market-based accounts.

Finally, studies indicate that extending the accounts to nonmarket consumption and investment would have a significant impact on estimates of income, product, and national wealth. Preliminary work on augmented accounting exclusive of the environment indicates that broadening the accounts to include comprehensive consumption and investment could easily double the reported net income and output and might increase reported net investment by a large factor.[3] Similarly, as discussed in Chapter 4, corrections for environmental flows, particularly those involving nonmarket impacts on the health and safety of the population, could have major impacts on measured income.

Comprehensive Accounts Provide Much Useful Information for Public Policy and Private Decision Making

Environmental accounts would provide useful information for managing the nation's assets and for improving regulatory decisions. For example, improved natural-resource and environmental accounts can provide useful information on natural assets under federal management. Better information on the value of minerals on federal lands would be useful in determining appropriate royalty rates and leasing policies for resources not allocated through competitive auctions. Better information on the stumpage value of timber in national forests would be useful not only for accounting purposes, but also for better management of these forests and for decision making on the balance among timber harvesting, wilderness preservation, recreation, and other uses. Efforts to prevent overfishing have been hamstrung by the lack of reliable information on changes in fish stocks. In many of these cases, data are already collected by federal agencies, but incorporating these data into the consistent frame-

---

[3]In the nonenvironmental sectors, inclusion of the value of human capital formation and the value of unpaid work and leisure time increases national saving, investment, and output significantly. For example, in Eisner's Total Income System of Accounts, comprehensive net domestic capital accumulation in 1981 is estimated to be 479 percent (see Eisner, 1988:Table S.3) of BEA's estimate of net private domestic investment (Eisner, 1989:158). Similarly, total output including nonmarket activity is estimated to be approximately double or more BEA's net national product in the comprehensive accounts of Nordhaus and Tobin (1972), Kendrick (1987), and Jorgenson and Fraumeni (1987). See Eisner (1988:Table S.5) for comparisons of market and comprehensive income and saving measures.

work of a set of national accounts would help regularize their collection and ensure consistency over time and across sectors.

In the case of environmental resources such as air and water quality, a comprehensive set of environmental accounts would provide useful information on the economic returns the nation is reaping from its environmental investments. The contrast between private and public investments is instructive. When a private company makes an investment in an automobile factory or a power plant, company accounts can be used to estimate the economic costs and benefits of that investment. Yet although the nation has allocated more than $1 trillion to environmental, health, and safety investments over the last quarter-century, there is no comparable set of accounts by which to reckon the nation's returns to those investments. Improved environmental accounts would also provide essential information for sound benefit-cost analyses in regulatory decision making. One of the most serious weaknesses in the U.S. environmental database is the lack of comprehensive and reliable data on actual human exposures to major pollutants. Better information on physical emission trends, human exposures, and the economic impacts and damages from air and water pollution would be valuable for expanded accounting measures of productivity. Hence, both the underlying information and the aggregate dollar values in environmental accounts would provide essential information for ensuring that our environmental regulations pass an appropriate cost-benefit test.

Investing in Comprehensive Accounts Would Yield a High Economic Return for the Nation

The federal government currently makes a substantial investment in collecting, analyzing, and distributing statistical data on the nation's economy. This information is valuable in part simply because we are curious about ourselves as a nation. We want to know what we are producing and consuming, to compare ourselves with other nations, and to assess the trends in economic activities. But provision of statistical data is also an investment in a public good. Having more complete, accurate, and timely data on economic activity requires the resources and data-collection abilities of the government. Data for economic accounts will not be provided by the private sector both because the private sector does not have access to the full range of administrative data available to government agencies and because there is little private economic profit in gathering and providing comprehensive economic accounts. Yet while the federal government invests heavily in the collection and distribution of economic data, it has to date invested very little in providing comprehensive economic accounts. And while many in the private sector have attempted to construct such accounts, private researchers have neither the resources nor the data to do so. As a result, the United States today has no set of comprehensive economic accounts, public or private.

An investment in comprehensive economic accounts would benefit the nation because, as noted earlier, better information allows both the public and private sectors to make better decisions. In particular, improved data on the interaction between the economy and the natural environment would have substantial economic benefit for the nation. Many examples of such benefits can be cited. Here we mention but a few from different areas to suggest the range of benefits to be derived.

One important area in which environmental accounting has proved useful is productivity. Growth in productivity, measured as output per person-hour, declined sharply after 1973. One of the leading explanations for this decline was that increased health and safety regulations were imposing significant economic burdens on the nation's businesses. Preliminary versions of natural-resource and environmental accounts—particularly the estimates of pollution control and abatement expenditures prepared by BEA—were of great value for estimating the impact of regulation and pollution-control expenditures on productivity. It is now generally accepted among productivity specialists that environmental regulation is responsible for some of the slowdown in productivity growth; without the existing environmental accounts, it is doubtful whether such a clear understanding would have been possible. Similar studies have analyzed the effect of pollution controls on agriculture and on coastal waters.[4]

One of the most difficult problems in environmental policy has been comparison of the costs and benefits of environmental regulations. The nation invests substantial sums in cleaning its air, water, and land. These investments have yielded substantial benefits in the form of declining emissions of many pollutants and fewer violations of air quality standards. What is unclear at present is the extent to which the expenditures have produced commensurate economic benefits in terms of improved human health, higher crop yields, and reduced property damage. Recent studies indicate that there have been substantial net economic benefits from pollution control (see Chapter 4). But these studies have not provided sufficient detail to allow pollutant-by-pollutant or sector-by-sector estimates of costs and benefits. Improved accounting systems for the environment could help refine our estimates and regulatory tools so that our pollution control investments might be more effectively allocated. There are major stakes involved here. A 10 percent reduction in pollution

---

[4]Some of these applications are summarized by Gianessi and Peskin (1976).

control expenditures due to improved information would amount to more than $10 billion per year in efficiencies for the nation.

Another potentially valuable application of environmental accounting relates to management of the nation's public lands. The nation's forests, rangelands, and waters provide a broad spectrum of valuable economic services. The federal government today reaps substantial revenues from timber harvesting, mining, and leasing of rangelands. A better set of accounts would probably indicate that current leasing policies are providing substantial subsidies. Between May 1994 and September 1996, mining companies patented claims on federal lands with an estimated gross mineral value of $15.3 billion, yet the charge to lessees for these claims was only $19,190. Because the accounting for federal mineral values was incomplete, the full resource value is not currently estimated. Similar subsidies are found in timber and rangeland (see Council of Economic Advisers, 1997). Improved accounts would help decision makers estimate the value of such federal assets and set more realistic prices for leases and patents.

Another area in which comprehensive accounts would be of great benefit is assessment of the costs and benefits of measures to slow greenhouse warming. Under the Kyoto Protocol of December 1997, the United States has undertaken to reduce its greenhouse gas emissions by 7 percent in the 2008-2012 period relative to 1990 emissions. The reductions are to include not only reduced emissions from industrial sources, but also the reductions resulting from carbon sequestration in forests. A comprehensive set of physical and economic accounts would provide the information base needed to estimate the carbon sequestration in forests. Current estimates are that approximately 200 million tons per year of carbon is being accumulated in forests. The nation would save $20 billion annually if a comprehensive set of measures and accounts verified this level of sequestration, if this sequestration could be used to offset industrial emission reductions, and if those industrial emission reductions cost $100 per ton of carbon. This is one of the most dramatic examples of the benefits of establishing comprehensive nonmarket physical and economic accounts.

Economists have developed a new view of the role of data collection, in which data are valuable because they allow better decisions to be made by both the public and private sectors. For example, better weather forecasting allows farmers to harvest their crops so as to reduce damage from frost. Another area that has been intensively studied is the value of better information about the science and economics of climate change. Governments and private firms, such as oil and coal companies and electric utilities, must cope with the enormous uncertainties in this area. Many of these uncertainties result from inadequate accounting of the costs of emission reductions and the potential impact of climate change in nonmarket sectors. Recent studies have found that improved information in this area would have substantial value. For example, it is estimated that reducing uncertainties about the costs and damages of climate change by half over the next two decades would be worth more than $20 billion.[5] Clearly, as the United States and other countries grapple with the conflict between their international commitments and the domestic costs of emission reductions, improved information on the economic costs and benefits involved could greatly benefit the analysis.

Link Between National Income Accounting and Measures of Sustainable Income

In light of increasing environmental problems in many sectors, concerns have been raised about the sustainability of current patterns of economic activity in both developed and developing countries. What are the environmental and economic implications of continuing "business as usual"? Will the current path of population, energy use, and growth of human settlements do irreversible harm to the natural ecosystems and life-support systems of the earth? Are we headed for economic overshoot and collapse if we continue to rely on today's technologies? In short, is our economy on a sustainable path? Economists have developed measures of national income and output that incorporate notions of sustainability. This section describes how the current national accounts are related to measures of sustainable consumption. A more complete discussion is contained in Appendix A.

Measures of national income take two fundamentally different approaches—one based on the idea of current production and one based on sustainable consumption. Those who originally constructed national income accounts were understandably concerned with obtaining accurate production-based measures of national output and income because much of their work took place in the shadow of the Great Depression. In the production-based view, tracking current production is seen as critical because it allows governments to take measures to stabilize the business cycle. As noted earlier, production-based measures usually rely on Hicksian income, which is the standard definition of net domestic or national product used in the national income accounts of virtually all nations today. The concept is production based in the sense that production in a given period is measured at market prices.

While standard production-based measures of income are useful tools for measuring current production, they do not directly address concerns

---

[5]For an example of estimates of the value of information about the science and economics of global warming, see Nordhaus and Popp (1997:Table 4).

about the sustainability of current decisions. As suggested earlier, it is conventional in economic analyses to define *sustainable national income* as the maximum amount that can be consumed while ensuring that all future generations can have living standards that are at least as high as those of the current generation.[6] Economic welfare, in this view, consists of per capita consumption of goods and services, both market and nonmarket. Consumption includes market items such as food, shelter, and entertainment; it also includes nonmarket items such as home-cooked meals or camping.[7]

What is the relationship between current measures of national output, such as net domestic product (NDP), and sustainable income? One of the most surprising results of modern economic theory is the output-sustainability correspondence principle (see Appendix A). This principle holds that under idealized conditions, net national product and sustainable income are identical. More precisely, when population is constant, when the national accounts include all stocks of capital and other dynamic features that affect production, and when markets accurately capture the entire social value of economic activity, NDP is an appropriate measure of sustainable income. In other words, the sum of total consumption and net capital formation is equivalent to the maximum sustainable amount of per capita consumption an economy can maintain indefinitely. Under idealized conditions, then, extending the NIPA to include comprehensive measures of consumption and net investment would make output and income more accurate indexes of sustainable income.[8]

The output-sustainability correspondence is of fundamental importance for guiding decisions about the design of the NIPA. However, important practical and theoretical qualifications to this principle must be emphasized. Augmented NDP will fail to measure sustainable income

---

[6]It should be emphasized that the definition of sustainability used here is chosen because it is particularly appropriate in the context of designing comprehensive national income accounts. Literally dozens of definitions and approaches have been suggested, and others may be more appropriate in different contexts.

[7]The economic approach to sustainability excludes many important individual values and collective activities because economic measurements do not go beyond the boundary of what can be directly or indirectly denominated in monetary units. The approach also considers the level of average or per capita consumption today and in the future. This high level of aggregation masks a number of important ways of disaggregating the complex ensemble of environmental and economic activities: it does not distinguish among the different future generations; it overlooks the distribution of consumption among different groups within a country or among countries; and it does not deal with issues of risk.

[8]This proposition dates back to Weitzman (1976). For a recent comprehensive treatment of the subject, see Aronsson et al. (1997).

accurately (1) if the list of consumption and asset categories is incomplete, (2) if there are technological changes or similar processes that are not captured in investment data, (3) if there are revaluation effects not calculated in the accounts, or (4) if there are market imperfections such as imperfect foresight. The significance of each of these issues is discussed in Appendix A. Even though the conditions under which the correspondence principle applies are quite stringent, the basic insight is of great value for the designing of environmental accounts.

ALTERNATIVE APPROACHES TO ENVIRONMENTAL ACCOUNTING

General Issues in Environmental Accounting

Over the last quarter-century, official statistical agencies and individual researchers have responded to the deficiencies in current accounting approaches by developing alternative approaches and novel systems of accounts. The first approaches, by individual researchers, tested wholly new frameworks that often included major aspects of nonmarket activities. Later, official statistical agencies began to take incremental steps toward including some activities that are near-market in nature. The differences in approach generally reflect varying emphasis on the deficiencies discussed earlier, differing views on the functions of national accounting, and differences in what are considered the appropriate functions of official statistical agencies.

An important difference among alternative approaches is the relative importance assigned to economic as opposed to physical accounting. Many approaches emphasize the importance of developing economic accounts, which, as discussed in the last section, are useful for both scorekeeping and management. Other approaches emphasize physical indicators, stressing the development of detailed information on physical flows and human exposures and impacts. Such an approach would be emphasized, particularly by official agencies, when construction of economic aggregates depended heavily on controversial analytical methods and imputations and when collection and dissemination of objective data was the primary goal.

One major issue involved in decisions about how far to extend the boundary of augmented and environmental accounts concerns data quality. As the accounts move further away from the current market boundary line, the quality of the data becomes increasingly suspect, and the cost of obtaining the data becomes increasingly large. Such market or near-market data as volumes and values of petroleum reserves or timber stocks can be estimated with an accuracy reasonably comparable to that of market data. On the other hand, obtaining data on nonmarket assets such as fishing stocks or the value of breeding potential is likely to be significantly more expensive. The data become even more fragmentary as one moves toward including environmental activities that have public-goods characteristics, such as the value of lower concentrations of particulate matter or improved visibility. Additionally, valuation sometimes involves highly complex and controversial approaches, such as use of survey questionnaires in which respondents are asked to place dollar values on hypothetical environmental conditions (an approach commonly referred to as "contingent valuation"; see Chapter 4). While private scholars might be willing to use back-of-the-envelope, or even seat-of-the pants, approaches, official statistical agencies are more reluctant to compromise their reputations with controversial and unproven methodologies.

The following subsections review approaches that emphasize physical accounting, the development of comprehensive economic accounts that include nonmarket activity and environmental services, and proposed approaches to environmental accounting developed by the United Nations.

Physical Accounting

One way to improve our understanding of the interaction between the economy and the environment is to supplement the accounts with improved physical information. Important examples are information on the state of the environment (e.g., ambient pollution levels and forest cover), the status of natural resources (e.g., reserves and resources of petroleum and natural gas), and the impacts of changing environmental conditions on human and ecosystem health (e.g., human exposures to different pollutants or pH levels in lakes). Such information can be arranged in a formal material-flow accounting system, such as that developed by Ayres and Knesse (1969). As demonstrated by Leontief (1970), the integration of such information with conventional economic data can be made quite rigorous by supplementing conventional input-output analyses with data on the flows of environmental pollutants. Sophisticated versions of such input-output matrices have been generated by Duchin and colleagues at New York University (see, for example, Duchin and Lange, 1993). This approach can be developed into a comprehensive input-output system that integrates economic and physical environmental information. An important example is the National Accounting Matrix Including Environmental Accounts (NAMEA) developed by Keuning and colleagues in The Netherlands (see Keuning, 1993). Similar physical accounting systems exist in Norway and France.

Most physical accounting efforts do not embed the information in an input-output framework, but attempt to be more descriptive. These approaches take account of the environment by assembling large quantities of descriptive physical information, such as indicators of air and water quality, species counts, and area of forest cover. Typically, these informal accounting systems appear as national state-of-the-environment reports or in large physical environmental databases such as the STRESS system in Canada and similarly large databases maintained by several U.S. governmental agencies.

Physical accounting systems play an important role in accounting and policy formulation. They provide the underlying data for regulatory analysis and for development of the aggregates that underlie economic accounting. Moreover, they provide rich physical and intuitive measures of environmental impacts. At the same time, several factors complicate their use for policy purposes. First, the choice of appropriate physical units of measure is not obvious. Presumably, the units of measure should be relevant for some environmental-policy concern. Should a forest, for example, be measured in terms of its acreage, the volume of its timber, the variety of its biota (as evidenced by the number of available species), the stock of nontimber resources such as firewood and grasses, or the number of miles and acres of fishable waters? From the policy maker's point of view, the answer will depend on policy objectives: commercial timber management, firewood supply, recreational uses, erosion protection, species diversity, and so on. Additionally, when environmental assets have multiple uses, as in the case of forests, the units of the indicators are different (acres, cubic feet, number of species, cords of firewood, and miles of streams). The noncommensurate nature of the different attributes makes physical accounting rich in detail, but poor for making policy decisions and determining tradeoffs.

In all accounting systems, important questions relate to coverage, detail, and aggregation. In an effort to encompass the many policy issues involved, physical systems can easily become quite large and detailed. Of course, national accounting systems are also enormous data systems—but most of the vast data iceberg is under water, and only the monetary aggregates are visible in the published numbers. Indeed, large data systems are worth little for scorekeeping, modeling, or policy purposes unless they can be aggregated in such a way that they can be digested and understood. While economists often suggest that measures should be aggregated in terms of dollar values (or present values if there are streams of values over time), putting physical measures into a common unit of account often involves difficult valuation issues. In many cases (for example, protection of unique resources such as the wildness of Yellowstone or the visibility at the Grand Canyon), policy makers may be uncomfortable with aggregating these unique values and trading them off against mundane things such as guns or butter.

Aggregation of data in different physical units requires weighting the various measures in order to convert them to a common unit of account. Often this aggregation is accomplished by using a common physical unit of measure, such as weight, volume, or energy content. However, this approach is seldom sensible because the environmental impacts per unit of physical measure differ by orders of magnitude according to the substance and the pathway of human exposure. Compare, for example, the impact of 20 kilograms of plutonium and sulfur in different delivery vehicles.

The Dutch NAMEA system converts dissimilar pollutants to common units on the basis of their contribution to environmental themes such as global warming and acid rain. Thus, emissions of greenhouse gases today are commonly measured in terms of their CO_2 equivalent or global-warming potential. While this approach is often sensible, it embodies hidden assumptions that may be highly controversial on close scrutiny. For example, the conversion of greenhouse gas pollutants to a common unit requires detailed scientific knowledge about the relative contributions of different gases, and if the contribution is nonlinear (as is almost always the case), the aggregation will be inaccurate. Sometimes, the aggregation includes hidden economic assumptions. For example, the usual approach to aggregating greenhouse gases is to take their contribution to global warming over 100 or 200 years, but not to discount them; this approach is generally flawed and may lead to inappropriate decisions (Reilly and Richards, 1993).

The above examples suggest that physical indicators are subject to many of the same pitfalls and difficulties that plague economic measures of nonmarket and environment activities. Physical accounting systems are most valuable for policy and scorekeeping purposes when overall environmental objectives and targets are clearly established. When the physical systems are highly complex and heterogeneous and are less closely linked to policy objectives, physical accounting is less useful for policy or accounting purposes. It is essential to emphasize, however, that detailed physical information remains an essential component of both economic accounts and environmental policy making.

At present, the United States places less emphasis on developing a comprehensive set of environmental indicators than do many other nations, especially in Europe. Recently, the need for a set of policy-relevant and scientifically based environmental indicators has received high-level attention. This point has been emphasized by the President's Council on Sustainable Development and the National Performance Review.

The development of improved environmental indicators is an important priority if the United States is to enhance its ability to evaluate and analyze environmental trends and to understand the interaction between the environment and the economy. To be useful for both policy-making and accounting purposes, these indicators should be designed to measure variables close to the area of ultimate concern; they should recognize the heterogeneity of environmental activities and damages, avoiding where possible simple national averages and recognizing the great diversity of the United States, particularly in areas where thresholds and non–linearities are important; and they should be capable of estimation through suitably stratified sampling, rather than requiring comprehensive population counts or inventories. From the point of view of environmental accounting, enhancing the accounts in a manner that is scientifically and economically sound will require considerable improvement in the underlying physical data.

Development of Comprehensive Economic Accounts

Comprehensive Measures of Income and Output

One approach to developing comprehensive economic accounts, and the one with the longest history, is the construction of comprehensive measures of national income or output to supplement the conventional national economic accounts. Many of these efforts have been broad-based attempts to address the general issues raised by the national accounts as discussed above, while others have focused primarily on the natural-resource and environmental components of the accounts.

Early efforts were part of the broader movement to construct more meaningful and comprehensive measures of economic welfare. These studies usually redefined the central concept of economic activity by extending both "consumption" and "output" to encompass large portions of nonmarket activity, sometimes including environmental activities. The first example of this approach is the Measure of Economic Welfare (MEW) indicator developed by Nordhaus and Tobin (1972). They defined an entirely new measure of economic welfare that included major new components such as leisure, nonmarket work, and imputations for the services of government and consumer capital. This measure also excluded activities that do not contribute to economic welfare, for example, commuting costs and regrettable necessities such as military spending. In addition, it subtracted an estimate of the environmental disamenities associated with urban activities.

A similar approach relying heavily on the concepts behind the MEW, Net National Welfare (NNW), was developed by the Japanese government (Japan, Economic Council, 1973). A number of comprehensive measures of output for the United States, including many nonmarket activities and assets as well as environmental activities, were developed by Zolotas (1981), Kendrick (1987, 1996), and Eisner (1985), and by Jorgenson in association with Fraumeni (1987) and with Christensen and Jorgenson (1969, 1973).[9] In addition, those emphasizing the ecological approach to economics have developed comprehensive accounts and attempted to value natural ecosystems (see Daly and Cobb, 1989, 1994; Costanza et al., 1997).

Targeted Approaches to Environmental Accounting

Comprehensive approaches are useful supplements to the conventional national accounts in that they can sketch the evolution of broad measures of economic activity. However, because most of the comprehensive approaches to measuring national output and income treat natural-resource and environmental measures in a broad-brush fashion, they do not provide many of the important details about particular sectors, environmental activities and assets, and interactions between the environment and the economy. Over the last two decades, many studies have taken a more targeted approach to environmental accounts, focusing on how the national accounts would be modified to incorporate the environment and offering estimates of economic activity in sectors that provide services of natural resources. One approach that treats the natural-resource and environmental sectors in more detail and illustrates many of the major issues involved is that developed by Peskin (1989a, 1989b). This approach was originally designed for the Measurement of Economic and Social Performance Project (sponsored by the National Science Foundation) and was recently adopted for the Philippine Environmental and Natural Resources Accounting Project (ENRAP) (see Peskin, 1989a, 1989b, for a description of the basic framework).

The ENRAP system is based on the principle that natural resources (including air, water, and land) have economic value because they generate valuable goods and services. Some of these services are marketed, such as commercial timber from forests, and these services are already included in the conventional market economic accounts. But many other valuable services—such as those associated with recreation, drinking water, and waste disposal—are not marketed, even though they have significant economic value.

---

[9]For a comprehensive review of these analyses, see Eisner (1988).

The ENRAP accounting framework starts with the conventional economic accounts and the conventional distinction between sectoral inputs and outputs. Those nonmarketed services that are inputs or intermediate goods (such as erosion protection that enhances agricultural production or land and water that provide disposal services) are added to the input side of the accounts. Perhaps the most important modification is on the output side. Outputs include not only marketed outputs, but also nonmarketed goods and services that go to investment or final consumption. Outputs also include as a negative item the environmental damage resulting from pollution. From an analytical point of view, capital formation includes such items as net increases of natural capital in the form of changes in forest stock or mineral reserves and carbon sequestration by forests. Added final consumption includes the value of recreational services such as visits to national parks or recreational fishing and the value of changes in health status due to changes in air quality or drinking water.

In conventional accounting, the marginal value of a produced good or service equals its price. The ENRAP approach recognizes that in the absence of market prices, there is nothing to ensure that the value of a nonmarket service will equal the price of that service. Therefore, to obtain accounting balance with the introduction of nonmarket environmental services, ENRAP introduces a term on the input side of the accounts— "net environment benefit." This entry is defined as the sum of the values of environmental input services (such as waste disposal), plus the values of positive output services (such as recreation), less any negative social damages (such as pollution damage) arising from the use of environmental inputs.

Another set of studies has focused on deriving estimates of economic activity in sectors providing services of natural resources or the environment, with emphasis on mineral fuels and forests. An important set of studies in this area has been undertaken by Repetto and colleagues at the World Resources Institute (Repetto et al., 1989). The principal thrust of this effort is to modify the conventional net national income and output by deducting estimates of the value of the depletion of natural resources such as forests, mineral stocks, fish stocks, and soils. The rationale for this modification is to ensure that reproducible capital and natural capital receive comparable treatment in the computation of net investment, net output, and national income. Expanding the boundary of the accounts also allows a more comprehensive definition of national saving and national wealth by including natural resources of minerals and forests, along with land and reproducible capital, in the definition of assets. More recently, the World Bank (1997) has provided estimates of augmented wealth, national output, and saving for a large number of developed and developing countries.

Some Common Issues in Environmental Accounting

In principle, it is economically sound to adjust conventional national output and income measures for final nonmarket consumption provided by the environment and other activities, as well as for the net capital accumulation in nonmarket assets. Adding nonmarket consumption and investment will produce a more accurate measure of sustainable income. Two decades of work on environmental accounting has shown, however, that there are significant obstacles to the construction of accurate estimates of augmented national income and output. For practical accountants, the most daunting obstacles are empirical and data problems involved in estimating quantities of stocks and flows and providing monetary valuation; these problems were discussed briefly above. There are also conceptual issues. To illustrate the challenges involved in including nonmarket activities, we discuss three such issues here—treatment of depreciation, treatment of pollution abatement expenditures, and issues of valuation.

Depreciation. According to standard national accounting conventions, the value of an asset is its market value, which, under competitive conditions, will equal the present value of the net returns from that asset. In this report, depreciation (or depletion for subsoil assets) is defined as the expected change in the present value of the returns on an asset due to aging. Under this approach, changes in present value due to changes in expected interest rates, expected prices, or expected physical flows are called *revaluations.* These definitions are the same for environmental assets, but for those assets the physical flows include both public and private services and damages, such as value of drinking water and adverse health effects. From an economic point of view, depreciation depends on the expected decline in the value of the economic services of an asset and not solely on its physical condition. For example, a forest might increase in value even though it had a declining volume of timber production if there were an increase in production of other goods and services.[10]

Asset values and depreciation can be estimated by calculating the discounted present value of the stream of net returns, market and non–market, from environmental capital. This calculation is complicated because it requires estimating future returns, capital lifetimes, and discount

---

[10]The definition of depreciation used here follows that in Fraumeni (1997). It differs slightly from the definition applied by BEA in defining capital consumption for the national accounts.

rates. A number of simplified methods have been proposed to overcome these difficulties, many of which are reviewed in detail in the next two chapters. Particularly for nonreproducible and renewable assets, alternative approaches give quite different answers, so caution must be used when applying those approaches to environmental assets. In examples of the value of net investment in subsoil assets for the United States, the results sometimes have differing algebraic signs under different approaches. These results emphasize the difficulties of ensuring precision when moving beyond the traditional boundaries of the marketplace.

Pollution expenditure accounting. One early suggestion for improving the conventional economic accounts, especially with respect to their neglect of environmental deterioration, was to treat all pollution abatement expenditures as "intermediate" expenditures in the national accounts. As a consequence, pollution abatement investments, governmental municipal sewage treatment expenditures, and defensive consumer outlays for pollution control would not be counted in GDP. The idea of deducting environmental protection and similarly defensive expenditures from GDP has a long history, but has never been formally adopted by national accountants, presumably because of difficulties in drawing the line between defensive and nondefensive outlays (see Nordhaus and Tobin, 1972; Juster, 1973).

One way of rationalizing a proposal to treat pollution abatement expenditures as intermediate inputs is to assume that these expenditures are just what is necessary to keep the stock of environmental capital intact—that is, to assume that the expenditures are exactly sufficient to offset any pollution and other environmental degradation. While convenient, this assumption is unlikely to be realistic for any particular time or sector. It seems likely that until the early 1970s, environmental quality in many areas was deteriorating, while since then it has improved in many areas.

Additionally, current measures of pollution abatement are defective because they are poor estimates of the true cost of environmental regulation. For example, many plants met regulations under the original Clean Water Act without making *any* expenditures on abatement equipment. They simply ceased producing certain highly polluting products, such as bright, highly coated writing papers. Other examples include the elimination of certain oil-based paints and of leaded gasoline. In these cases, abatement expenditures were zero, although true economic costs were positive. In other cases, abatement expenditures overstate true opportunity costs because of the difficulty of separating accounting costs into pollution abatement and other costs. Moreover, many pollution abatement activities are voluntary and are not in response to policy. For example, a major component of the U.S. pollution abatement expenditure series is expenditures associated with sewer hookups and septic tanks for newly constructed housing. Such sanitary practices have a history that long predates the environmental movement, and recent activity reflects local laws, building codes, and zoning ordinances regarding pollution. If expenditures associated with such conventional practices are not excluded from pollution abatement expenditure series, the use of such series to estimate the costs of regulation or explain productivity changes can yield very misleading results. How large might this overestimate be? One study has shown that about 20 percent of reported pollution abatement expenditures in the United States did not originate in federal regulatory policy. In some industrial sectors, nearly all reported expenditures predated federal regulations.[11]

Valuation. Traditional economic accounts use market prices to value intermediate and final output. For near-market or nonmarket activities, economists rely on alternative approaches that use proxies for market prices or develop alternative methods that impute values indirectly. One example is firewood collected by households on their own property; here, the appropriate value would be the market price of equivalent commercially sold firewood. Life becomes more complicated when the goods and services have no market equivalent or have public-goods characteristics. In these cases, values are often imputed (1) by using surveys, (2) by unbundling the commodities and valuing component parts, or (3) by looking at behavior that reveals consumer valuation of the commodities. These three techniques are exemplified by contingent-valuation surveys, hedonic regressions, and the travel-cost method, respectively.[12]

A key feature of the appropriate design of augmented accounts is that prices or values should always be measured by the value of the *marginal* or *last* unit of the good or service consumed. That is, the value of bottled water is not the average value, but the value of the last unit drunk, which will be significantly lower than the average; the difference between average and marginal value is called consumer surplus. This convention of using the marginal value is employed throughout the national economic accounts, so adhering to this approach will ensure that environmental goods and services are valued consistently with market goods and services. One of the difficulties in adopting valuations from the existing

---

[11]Much of this discussion draws on examples from Gianessi and Peskin (1976).

[12]A useful survey of valuation issues for environmental resources is provided in Smith (1996).

environmental literature is that many valuation studies calculate the average rather than the marginal values (that is, they add in consumer surplus, which is appropriate for economic accounts). Use of average values will tend to overstate the total value of an item relative to market goods and services.

There is an interesting parallel here between valuation issues for environmental services and difficulties in measuring the cost of living. Recently, a group of economists undertook a comprehensive assessment of the adequacy of the Consumer Price Index (CPI) (Advisory Commission to Study the Consumer Price Index, 1996). The commission collected a wide variety of studies and investigated whether the CPI accurately measures the trend in the cost of living. The commission concluded that the CPI has a significant upward bias, primarily because of an inadequate treatment of quality change.

Criticisms of the CPI revolve primarily around the difficulty of measuring nonmarket services. That is, a major criticism of the CPI is that it measures the prices of the market goods that consumers purchase rather than the prices of the nonmarket services these goods deliver. Thus the CPI measures the prices of automobiles, electricity, and hospital days, not the costs of travel, lighting, or delivering a baby. To estimate the service prices for consumer purchases would require—in a way directly parallel to valuation of environmental goods and services—imputing or calculating the values of nonmarket services. There are no deep theoretical problems involved in the estimation of these nonmarket values, but they present measurement problems in practice because transaction prices for the services are almost never observed. Many of the difficulties that arise in correcting the CPI for quality change are analogous to the difficulties that arise in valuing nonmarket environmental goods and services. Indeed, correcting the CPI for quality change would probably be easier than estimating the proper values of environmental goods and services because the CPI includes primarily private goods, while valuation of environmental services involves public goods as well.

Extensions of the U.N. System of National Accounts: System of Integrated Environmental and Economic Accounting

A great deal of work outside the United States has been devoted to developing physical and monetary accounts for natural resources and the environment (see for example Uno and Bartelmus, 1998). Because of the diversity of approaches and controversies about alternative methodologies, however, no international consensus has been reached on the appropriate model for establishing a uniform system of environmental accounts. Therefore, it was decided for the 1993 SNA to treat environmental accounts as satellite accounts. Environmental accounts would thereby serve as a tool for expanding the analytical capacities of the national accounts without changing the core accounts, thus complementing rather than substituting for the traditional national accounts (see United Nations, 1984, 1991, 1993).

The various approaches were compiled and synthesized in the United Nations System of Integrated Environmental and Economic Accounting (SEEA) (United Nations, 1993). Unlike the SNA, the SEEA has not been adopted as an international standard and should be viewed as a set of proposals for environmental accounts.

The SEEA is a highly flexible framework encompassing approaches that range from reorganizing the current accounts to building a full set of household and nonmarket service accounts. The basic framework envisions adding environmental flows in a series of steps or versions. Version I of the SEEA reorganizes the traditional national accounts to highlight environmental and natural-resource flows. Version II is a restatement of the expenditure-accounting approaches describing the monetary and physical flows and stocks. Version III links the physical information of version II with the monetary data of version I. Version IV imputes environmental damages to obtain a more comprehensive measure of output and includes the depletion of natural resources and environmental pollution costs. Version V, which has not been extensively discussed, considers more radical extensions, such as extending the production boundary in the household sector and introducing environmental services as an output.

Since version IV has received the most international attention, it is the focus of the discussion here. Version IV treats environmental degradation and depletion as subtractions from net product. In effect, both depletion and degradation are viewed as sources of depreciation of natural capital. We focus here on two examples of how the SEEA differs from alternative approaches. One is in the estimation of depletion of natural resources such as petroleum, which is valued at market values or sometimes at replacement cost. A second is the cost of environmental degradation—such as water pollution—which is treated either as "costs caused" or "costs borne." Under this distinction, degradation can be valued in terms of either the costs to the sector if it were to eliminate the degradation (costs caused) or the damage to producing or affected sectors due to the degradation (costs borne). For the most part, when implementing the SEEA, researchers have relied on the costs-caused approach. These depletion and degradation estimates are subtracted from conventionally defined value-added to derive environmentally adjusted net income measures.

It is useful to highlight the fact that the SEEA relies heavily on costs in its design of environmental accounts. Although it is common practice today, the use of restoration-cost estimates to measure environmental degradation and replacement cost to measure natural-resource depreciation is an inconsistent and inappropriate practice. The appropriate approach is to measure the market value (along with the relevant value of nonmarket impacts) of any change in the services of these environmental assets and of the change in the stocks of these assets. As is discussed in Chapter 3, for example, the appropriate valuation of depletion of petroleum stocks is the change in the market value of oil in the ground.

Use of the SEEA methodology can lead to inappropriate results. Suppose that the environment is initially clean and that the market and nonmarket damages from emitting a few grams of dust are very small. The cost of maintaining the clear environment by reducing those last few grams of dust might be enormous. Thus, the use of restoration costs as a measure of pollution control benefits (or damages) can lead to a significant overestimate of benefits. Paradoxically, if restoration costs are used to measure damage, the clean economy may be shown to be more environmentally damaged than the dirty society.[13] With respect to the degradation estimates, the authors of the SEEA recognize the theoretical difficulties involved in using cost-caused or replacement-cost data. While leaving the door open for the use of valuation estimates based on damages or costs borne, the authors are skeptical of the practical use of valuation techniques and of similar imputation methods.

A second notable feature of the SEEA framework is its adherence to conventional SNA sectoring or production boundaries. The close adherence to SNA concepts is an important advantage since it helps ensure consistency with the core accounts. Also, since the SNA framework has been widely adopted, the close adherence of the SEEA to the SNA will help ensure international comparability. This consistency comes at a price, however. The most important shortcoming arises because of the omission of nonmarket services and investments. There is no place in the SEEA system (at least with versions I through IV) for the amenities provided by the environment in the form of recreation, health impacts, erosion control, or disposal services. The SEEA in effect equates the term "nonmarket" with "noneconomic."

As a result of the omission of environmental services, the economic link between the economic value of an environmental asset and the services it provides is broken. Thus, while a forest can have economic value

---

[13]An example of the difficulties is seen in the environmental adjustments for the United States using the SEEA approach in Grambsch and Michaels (1994). These adjustments are quite large (about 8 percent of GDP), primarily because of the use of restoration costs to measure environmental damages.

in the SEEA framework, this value comes only from the commercial products of the forest, such as timber, and not from other forest services, such as watershed protection, recreational services, and carbon sequestration. By neglecting nonmarket assets and services, the SEEA also limits the coverage of household production. Some household production—such as the production of nonmarketed firewood—has both substantial economic value to households and serious environmental consequences due to the pollution from the smoke.

A third key feature of the SEEA is the treatment of natural-resource depletion. In conventional accounting, net investment is measured as the change in the value of the stock of an asset between two periods. The SEEA does not employ the usual definition of net investment, but focuses only on natural-resource depletion. Under the SEEA, when resources are depleted, there is a deduction from net output; but when resources are discovered, there is no increment to net output. Hence, even though the stock of petroleum reserves is constant over time, the SEEA would be recording a series of deductions from output and income to reflect petroleum production. The SEEA logic is that discovered resources are not really additions to the stock; they merely represent a shift from the nonproduced, noneconomic stock of assets to the nonproduced, economic stock. If, however, the stock of petroleum were valued in terms of its market value or discounted services, additions and depletions would be treated more symmetrically. Discoveries would increase the value of the stock, while depletion would decrease its value.

Environmental Accounting in Other Countries

As noted earlier, environmental and natural-resource accounting has been extensively developed in countries outside the United States over the last quarter-century. As in the United States, these augmented accounts represent an attempt to cast light on the interactions between the economy and the environment. In other industrialized countries, three main areas of concern have been identified:

• *Depletion*—Some countries have been concerned about the depletion of scarce natural resources. Particularly in northern Europe, where North Sea oil and gas resources constitute a significant fraction of natural assets, policy makers want to determine the extent to which nonrenewable resources are being depleted.

• *Degradation*—Many countries have been concerned about the degradation of the natural environment through pollution. Pollution not only renders the air, water, and soils less productive, but has health impacts and degrades people's enjoyment of the environment.

• *Protection*—Countries adopt numerous measures to protect or re-store the environment. These activities include pollution abatement and control expenditures, research into cleaner technologies, ecological taxes, and fiscal incentives for environmentally benign production patterns. Policy makers are interested in the economic costs and impacts of such environmental protection measures.

Each of these three major facets of the interaction between the economy and the environment corresponds to a different aspect of economic policy and requires different data sets, concepts, and classifications. A comprehensive environmental accounting system addresses each of these sets of issues. Environmental accounting in the United States has to date addressed primarily the issue of mineral depletion and (up to 1995) the cost of environmental protection. Approaches being applied in other countries involve analyzing different parts of the interaction, such as material flows into and within the economy; recycling; the costs of meeting environmental targets; ecological taxes; and emissions of various pollutants into the air, water, and soils.

As noted earlier, other countries have adopted BEA's approach of keeping their environmental and natural-resource accounts in satellite or supplemental accounts; the core national accounts have not been modified to reflect environmental and natural-resource changes. This approach has been endorsed by the European Commission (European Union, 1994:5):

> The development of a "greened" GNP, although having a certain appeal . . ., raises a number of difficult methodological questions which rule it out as a realistic option for the foreseeable future. Therefore what is needed—as a first step—is an approach which makes environmentally interesting parts like resource depletion and environment degradation, firstly in the form of physical indicators, later with the help of available techniques transformed into monetary value, still—however—keeping the various building blocks of such a system of integrated environmental and economic accounting separate, a so-called satellite approach.

Thus the European Union has decided to take the same approach as the United States: to focus on creating multiple, integrated data sets that track the interaction between the economy and the environment. This approach emphasizes the multidimensional nature of the interaction, rather than attempting to create a single-number modified GDP.

Table 2-1 provides a summary of the major sectors that have been studied in environmental accounts of various high-income countries. This table refers to published studies by official statistical agencies comparable

TABLE 2-1 Development of Environmental and Natural-Resource Accounts in Major Industrial Countries (synoptic table illustrating areas in which countries are working)

Country	Natural-Resource Accounts				Material Flows	Emissions	Pollution Abatement/ Control
	Forests	Subsoil Assets	Water	Land			
Australia	x	x		x			x
Austria	x			x	x	x	x
Belgium							
Canada	x	x	x				x
Denmark	x	x	x	x		x	
Finland	x		x		x	x	x
France		x	x	x		x	x
Germany	x			x	x	x	x
Holland		x	x		x	x	x
Italy			x			x	x
Norway	x	x			x	x	x
Spain			x		x		
Sweden	x				x	x	x
United Kingdom	x	x	x	x		x	x
United States		x		x			x

Note: The x's in this table are indicative only. Work is at different stages of maturity, and the situation changes rapidly.

to BEA. In addition, extensive work in other areas has been undertaken by private research institutes. To a considerable extent, the focus of the environmental and natural-resource accounting of each country reflects its own national priorities and policy concerns. Therefore, Canada and the Scandinavian countries have highlighted forestry accounts, while densely populated Holland has focused more intensively on pollution of air, rivers (particularly the Rhine), and soils.

Some of the environmental accounts are quite close to the existing national accounts; this is particularly the case for the mineral accounts, which are conceptually included in existing national wealth accounts. Other accounts consist primarily of disaggregating existing transactions, such as those concerned with pollution abatement and control expenditures. Another set of accounts represents an extension of existing input-output systems to include physical flows of pollutants along with the purchases and sales of goods and services.

The above review indicates that the principles and practices of environmental and natural-resource accounting are well developed in major industrial countries. Countries are concerned about the interaction between the economy and the environment, particularly as regards the extent of resource depletion and environmental degradation, as well as the economic costs of environmental protection. Other countries follow the U.S. practice of analyzing environmental linkages in satellite accounts, which provide useful data for both management and scorekeeping without changing the core national economic accounts.

U.S. INTEGRATED ENVIRONMENTAL AND ECONOMIC SATELLITE ACCOUNTS

History of Environmental Accounting in the Commerce Department

Many of the issues considered in the current discussion about expanding the traditional NIPA were involved in the earliest decisions about designing the accounting framework. From the beginning, those within BEA who constructed the NIPA considered aspects of what is now called environmental accounting. It was decided at the outset to focus primarily on an accounting framework whose boundary encompassed market transactions. Interestingly, in early efforts, depletion of mineral assets was a deduction from national product for obtaining net output. This practice was discontinued and depletion removed because the approach was thought to be asymmetrical in subtracting depletion without adding additions.

As the idea of augmented accounting began to emerge in the early 1970s, BEA came under pressure to expand its accounts to include significant nonmarket activities, with an eye to improving the accuracy of the accounts as a measure of economic well-being. At that time, BEA was not inclined to develop augmented accounts because it believed that imputations of the volume and values of nonmarket activities would be subjective and based on unproven methodologies and would lead to a deterioration in the accuracy of the national accounts.

BEA's initial concern was with the failure of the accounts to treat pollution abatement expenditures consistently. Specifically, if some part of final consumption were devoted to pollution control, GDP would not be affected, but if the business sector devoted the same level of resources to pollution abatement, conventionally measured GDP would fall. Rather than making major conceptual changes to the national accounts, BEA recommended that a separate series on pollution abatement expenditures be developed to interpret movements in national output, rather than to change the definition of national output itself. Work began on the development of such a survey in 1971, and preliminary results were published in 1973. Ironically, this first foray of BEA into environmental accounting was eliminated in the budget cuts of the 1990s.

In 1972, Congress directed the Secretary of Commerce to study the effect of the costs of the Clean Water Act on manufacturers. In response, BEA and the Bureau of the Census developed an environmental statistics program. BEA formed an Environmental Studies Staff within the Office of the Director. Several activities were initiated to support the planned pollution abatement expenditure series. In particular, BEA added a number of questions on pollution abatement to the November 1973 Plant and Equipment Survey of companies, and in 1974 the Census Bureau began surveying about 19,000 manufacturing establishments with regard to their pollution abatement expenditures. In response to the success of BEA's efforts to develop pollution abatement expenditure data, Congress approved funds that allowed for expansion of BEA's environmental program. In 1977, BEA established the Nonmarket Economics Division, which consisting of three branches—the Abatement and Control Expenditures Branch, the Unit Costs and Emissions Branch, and the Measures of Economic Well-Being Branch (Bureau of Economic Analysis, 1987).

The first two of these branches were involved primarily in reconfiguring the data on costs that were already contained in the national economic accounts. The Measures of Economic Well-Being Branch—in a significant departure from conventional national income accounting—focused on the deficiencies of conventional income and output measures as measures of social well-being. The Economic Well-Being Branch conducted research on a number of issues that would arise if the traditional accounts were expanded to better reflect societal well-being. These studies involved the value of nonmarketed household work, the value of services associated with governmental capital and consumer durables, the investment value of education and training, and the value of the discovery and depletion of minerals (Bureau of Economic Analysis, 1982). Methodologies developed in the work on minerals accounting contributed directly to the development of the mineral accounts in Phase I of BEA's IEESA.

Although interest in environmental and resource accounting was growing outside the United States, the Measures of Economic Well-Being Branch was abolished in 1981. The Nonmarket Economics Division was cut and renamed the Environmental Economics Division; the work of this new division was confined to the generation and analysis of pollution abatement expenditure data. In 1995, the balance of the program, along with the pollution abatement expenditure survey, was abolished to meet the budget cuts of the 1990s. In the meantime, BEA initiated work in the IEESA system in 1992, but this work, as noted earlier, was stopped by Congress in 1994.

Overview of the Integrated Environmental and
Economic Satellite Accounts

Work on the IEESA began in earnest in 1992 and was accelerated when President Clinton emphasized its importance in his Earth Day speech in 1993. The essential framework for the IEESA, along with a proposed framework for future study, was set forth in two articles in 1994 (see Bureau of Economic Analysis, 1994a, 1994b). As envisioned by BEA at that time, the work plan would have three phases:

• *Phase I, Overall Framework and Prototype Estimates for Subsoil Assets*—The first phase involved establishing the overall framework and process for developing prototype satellite accounts for subsoil assets such as oil, gas, and nonfuel minerals. The focus was "on proved reserves, the basis for valuation is market values, and the treatment given mineral resources—which require expenditures to prove and which provide 'services' over a long timespan—is similar to the treatment of fixed capital in the existing accounts" (Bureau of Economic Analysis, 1994a:48-49). The Phase I report of these two articles presented a preliminary view of the framework of U.S. environmental accounts, along with numerical estimates of the values of additions, depletion, and stocks for major subsoil mineral resources.

• *Phase II, Renewable Natural Resources*—The second phase "calls for work to extend the accounts to renewable natural resource assets, such as trees on timberland, fish stocks, and water resources" (Bureau of Economic Analysis, 1994a:49).

• *Phase III, Environmental Assets*—The third phase "calls for moving on to issues associated with a broader range of environmental assets, including the economic value of the degradation of clear air and water or the value of recreational assets such as lakes and national forests" (Bureau of Economic Analysis, 1994a:49).

Since publishing its first report in 1994 in the two above-mentioned articles, BEA has ceased further work on environmental accounting in response to the congressional stop-work order.

SUMMARY AND CONCLUSIONS

The last quarter-century has seen an increasing awareness of the interactions between human societies and the natural environment in which they thrive and upon which they depend. This awareness has been dramatically heightened by concerns about resource scarcity, environmental degradation, global environmental issues, and the possibility that the economy is not sustainable. The combination of this increasing awareness and recognition of the primitive state of environmental data has led to a widespread desire to broaden the nation's economic accounts to include natural resources and the environment. The idea of including natural-resource and environmental assets and services in the economic accounts is part of a movement to develop broader economic indicators. It reflects the reality that economic and social welfare do not stop at the market's border, but extend to many near-market and nonmarket activities.

BEA has studied augmented accounting since the early 1980s. It began work on the U.S. version of environmental accounting, the IEESA, in 1992. Congressional concerns about environmental accounting were raised shortly after the first publication of the U.S. environmental accounts, and Congress requested that work on the IEESA cease until the methodological issues had been reviewed. In response to the congressional mandate, the Commerce Department asked the National Academy of Sciences to undertake a review of environmental accounting, and this report is a response to that request.

The NIPA are the most important measures of a country's overall economic activity. From the perspective of environmental accounting, the major point to recognize is that GDP is conceptually defined to include the final output of marketed goods and services—that is, goods and services that are bought and sold in market transactions. While recognizing the need to consider alternative measures, it is important to retain the core market-based accounts, which are of great value for historical and international comparisons and will continue to be a critical indicator for much economic policy making.

Work on augmented accounting in official statistical agencies, as well as by individual scholars, has yielded estimates on a wide variety of nonmarket activities for experimental augmented national accounts. The guiding principle in extending the national economic accounts is to measure as much economic activity as feasible, regardless of whether it takes place inside or outside the marketplace. Augmented national economic accounts are designed to provide better measures of final output—of what consumers in the United States currently enjoy in the way of goods and services, and of the accumulation of capital of all kinds that will permit the future production of goods and services.

A set of well-designed environmental accounts can overcome the recognized shortcomings of the current market-based accounts. They can provide useful information for managing the nation's public and private assets, for improving regulatory decisions, and for informing private-sector decisions. The collection of data on comprehensive income and output is an investment that would have a high economic return for the nation. There are many examples of the benefits of comprehensive economic accounts. These include better estimates of the impact of regulatory programs on productivity, analyses of the costs and benefits of environmental regulations, management of the nation's public lands and resources, and assessment of the costs and benefits of taking steps to slow global warming.

Augmented national accounts can also provide valuable indicators of whether economic activity is sustainable. The national accounts have a close relationship with measures of sustainable income, since the usual measure of NDP corresponds to the highest sustainable level of per capita consumption under idealized conditions.

The nation's measures of national income and output can be improved by including all consumption and net investment to obtain augmented income and output measures. Among the currently omitted items that need to be added are nonmarket consumption, such as home production and final environmental services, and nonmarket investments, such as changes in the value of resource stocks and investment in human capital.

Over the last quarter-century, official statistical agencies and individual researchers in the United States and abroad have responded to the deficiencies in current accounting approaches by developing alternative approaches and new systems of accounts. An initial general approach is to supplement the accounts with improved data on physical flows. Physical accounting systems are valuable for policy purposes when overall environmental objectives and targets are clearly established. They are an essential component of both economic accounts and environmental policy making. At present, the United States has invested little in developing comprehensive environmental indicators. The development of improved environmental indicators is an important priority for enhancing the nation's ability to evaluate and analyze environmental trends and track the interaction between the environment and the economy. From the point of view of environmental accounting, enhancing the national accounts in a manner that is scientifically and economically sound will require considerable improvement in the underlying physical data.

A second approach to environmental accounting is the construction of comprehensive measures of national income or output to supplement the conventional GDP and NDP accounts. Many efforts in this area have been broad-based attempts to remedy general issues raised by the national accounts. Other studies have introduced augmented environmental accounts with a more targeted approach, focusing on how the national accounts would be modified to incorporate the environment and offering estimates of economic activity in sectors providing services of natural resources.

Because of the diversity of approaches and controversies about appropriate approaches, no international consensus has been achieved on a uniform system of environmental accounts. The 1993 SNA entailed developing environmental satellite accounts as a way of expanding the analytical capabilities of the national accounts without changing the core accounts. In this proposal, environmental accounts would complement rather than substitute for traditional accounts.

The principles and practices of environmental and natural-resource accounting are well developed. Countries are concerned about the interaction between the economy and the environment, particularly the extent of resource depletion and environmental degradation, as well as the economic costs of environmental protection. Other countries follow the U.S. practice of analyzing environmental linkages in satellite accounts, which provide useful data for both management and scorekeeping without changing the core NIPA.

Intensive work on the IEESA began in the United States in 1992. As envisioned by BEA, the work plan would have three phases: Phase I, which involved establishing the overall framework and developing prototype satellite accounts for subsoil assets such as petroleum, gas, and nonfuel minerals; Phase II, which would extend the accounts to renewable natural-resource assets, such as trees on timberland, fish stocks, and water resources; and Phase III, which would extend the effort to issues associated with a broader range of environmental assets, including the economic value of the degradation of clear air and water and the value of recreational assets such as lakes and national forests.

3

Accounting for
Subsoil Mineral Resources

INTRODUCTION

Subsoil minerals—particularly petroleum, natural gas, and coal—have played a key role in the American economy over the last century. They are important industries in themselves, but they also are crucial inputs into every sector of the economy, from the family automobile to military jets. In recent years, the energy sector has been an important contributor to many environmental problems, and the use of fossil fuels is high on the list of concerns about greenhouse warming.

The National Income and Product Accounts (NIPA) currently contain estimates of the production of mineral products and their flows through the economy. But the values of and changes in the stocks of subsoil assets are currently omitted from the NIPA. The current treatment of these resources leads to major anomalies and inaccuracies in the accounts. For example, both exploration and research and development generate new subsoil mineral assets just as investment creates new produced capital assets. Similarly, the extraction of mineral deposits results in the depletion of subsoil assets just as use and time cause produced capital assets to depreciate. The NIPA include the accumulation and depreciation of capital assets, but they do not consider the generation and depletion of subsoil assets.

The omission is troubling. Mineral resources, like labor, capital, and intermediate goods, are basic inputs in the production of many goods and services. The production of mineral resources is no different from the production of consumer goods and capital goods. Therefore, economic accounts that fail to include mineral assets may seriously misrepresent trends in national income and wealth over time.

Omission of minerals is just one of the issues addressed in the construction of environmental accounts. Still, extending the NIPA to include minerals is a natural starting point for the project of environmental accounting. These assets—which include notably petroleum, natural gas, coal, and nonfuel minerals—are already part of the market economy and have important links to environmental policy. Indeed, production from these assets is already included in the nation's gross domestic product (GDP). Mining is a significant segment of the nation's output; gross output originating in mining totaled $90 billion, or 1.3 percent of GDP, in 1994. This figure masks the importance of production of subsoil minerals in certain respects, however, for they are intimately linked to many serious environmental problems. Much air pollution and the preponderance of emissions of greenhouse gases are derived directly or indirectly from the combustion of fossil fuels—a linkage that is explored further in the next chapter. Moreover, while the value of mineral assets may be a small fraction of the nation's total assets, subsoil assets account for a large proportion of the assets of certain regions of the country.

Current treatment of subsoil assets in the U.S. national economic accounts has three major limitations. First, there is no entry for additions to the stock of subsoil assets in the production or asset accounts. This omission is anomalous because businesses expend significant amounts of resources on discovering or proving reserves for future use. Second, there is no entry for the using up of the stock of subsoil assets in the production or asset accounts. When the stock of a valuable resource declines over time through intensive exploitation, this trend should be recognized in the economic accounts: if it is becoming increasingly expensive to extract the subsoil minerals necessary for economic production, the nation's sustainable production will be lowered. Third, there is no entry for the contribution of subsoil assets to current production in the production accounts. The contribution of subsoil assets is currently recorded as a return to other assets, primarily as a return to capital.

There is a well-developed literature in economics and accounting with regard to the appropriate treatment of mineral resources. The major difficulty for the national accounts has been the lack of adequate data on the quantities and transaction prices of mineral resources. Unlike new capital goods such as houses or computers, additions to mineral reserves are not generally reflected in market transactions, but are determined from internal and often proprietary data on mineral resources. Moreover, there are insufficient data on the transactions of mineral resources, and because these resources are quite heterogenous, extrapolating from existing transactions to the universe of reserves or resources is questionable.

Notwithstanding the difficulties that arise in constructing mineral accounts, the Bureau of Economic Analysis (BEA) decided this was the best place to begin development of its Integrated Environmental and Economic Satellite Accounts (IEESA). BEA in the United States and comparable agencies in other countries have in recent years developed satellite accounts that explicitly identify mineral assets, along with the changes in these assets over time. This chapter analyzes general issues involved in minerals accounting and assesses the approach taken by BEA (as described in Bureau of Economic Analysis [1994b]). The first section provides an overview of the nature of subsoil mineral resources and describes the basic techniques for valuing subsoil assets. The second section describes BEA's approach to valuation, including the five different methods it uses to value subsoil mineral assets. The third section highlights the specific strengths and weaknesses of BEA's approach, while the fourth considers other possible approaches. The chapter ends with conclusions and recommendations regarding future efforts to incorporate subsoil mineral assets into the national economic accounts.

GENERAL ISSUES IN ACCOUNTING FOR MINERAL RESOURCES

Basics of Minerals Economics

A mineral resource is "a concentration of naturally occurring solid, liquid, or gaseous material, in or on the earth's crust, in such form and amount that economic extraction of a commodity from the concentration is currently or potentially feasible" (Craig et al., 1988:20). The size and nature of many mineral resources are well known, whereas others are undiscovered and totally unknown. Figure 3-1 shows a spectrum of resources that differ in their degree of certainty, commonly described as measured, indicated, inferred, hypothetical, and speculative. Another important dimension is the economic feasibility or cost of extracting and using the resources. Some resources are currently profitable to exploit; others may be economical in the future, but currently are not. Along this dimension, mineral resources are conventionally described as economic (profitable today), marginally economic, subeconomic, and other.

Resources that are both currently profitable to exploit (economic) and known with considerable certainty (measured or indicated) are called reserves (or ores when referring to metal deposits). This means reserves are always resources, though not all resources are reserves.[1]

---

[1]Two additional categories of mineral endowment are worth noting since they are commonly encountered. The reserve base encompasses the categories of reserves and marginal reserves, as well as part of the category of demonstrated subeconomic resources shown in Figure 3-1. While reserves and the reserve base are typically a small subset of resources,

62

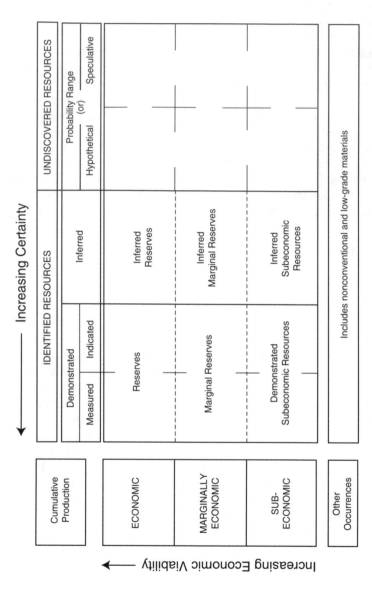

FIGURE 3-1 Classification of Mineral Resources. Source: Mineral Commodities Summaries, U.S. Geological Survey (1992:203).

Over time, reserves may increase. Exploration may result in the discovery of previously unknown deposits or demonstrate that a known deposit is larger than formerly indicated. Research and development may produce new techniques that allow previously known but uneconomic resources to be profitably extracted. A rise in a mineral commodity's price may also increase reserves by making previously unprofitable resources economic.

The exploration required to convert resources into reserves entails a cost. As a result, companies have an incentive to invest in the generation of new reserves only up to the point at which reserves are adequate for current production plans. For many mineral commodities, therefore, reserves as a multiple of current extraction tend to remain fairly stable over time.

While by definition all reserves can be exploited profitably, the costs of extraction, processing, and marketing, even for reserves of the same mineral commodity, may vary greatly as a result of the reserves' heterogenous nature. Deposit depth, presence of valuable byproducts or costly impurities, mineralogical characteristics, and access to markets and infrastructure (such as deepwater ports) are some of the more important factors that give rise to cost differences among reserves.

Figure 3-2 reflects the heterogenous nature of mineral resources by separating the reserves and other known resources for a particular mineral commodity according to their exploitation costs.[2] The lowest-cost reserves are in class A; their quantity is indicated in the figure as 0A and their exploitation costs as $0C_1$. The next least costly reserves are found in class B, with a quantity of AB and a cost of $0C_2$. The most expensive reserves are found in class M. These reserves are marginally profitable. The market price 0P just covers the extraction cost of class M ($0C_m$) plus the opportunity cost (C_mP) of using these reserves now rather than saving them for future use. This opportunity cost, which economists refer to as *Hotelling rent* (or sometimes scarcity rent or user cost) is the present value of the additional profit that would be earned by exploiting these reserves at the most profitable time in the future rather than now.[3]

---

resources in turn are a small subset of the resource base. The resource base, not illustrated in Figure 3-1, encompasses all of a mineral commodity found in the earth's crust.

[2]Similar comparative cost curves are used to illustrate the relative costs of mineral production for major producing countries or companies. See, for example, Bureau of Mines (1987) and Torries (1988, 1995).

[3]Where the relevant market for a mineral commodity is global and transportation costs are negligible, Figure 3-2 reflects cost classes for reserves and other known resources throughout the world. Where a mineral commodity is sold in regional markets, a separate figure would be required for each regional market, and the cost classes shown in any particular figure are only for the reserves and other known resources in the regional market portrayed.

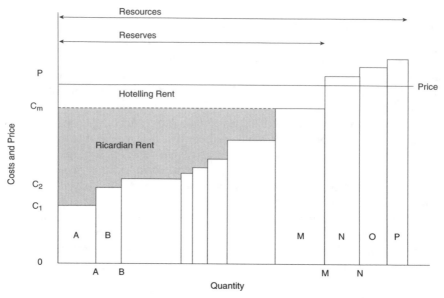

FIGURE 3-2   Mineral Reserves and Other Known Resources by Cost Class.

Known resources in Figure 3-2 with costs above those of class M, such as those in classes N, O, and P, are by convention not reserves. In this case, mineral producers, like other competitive firms, will have an incentive to produce up to the point where the current production costs of the next unit of output, inclusive of rents, just equals the market price. When Hotelling rents exist, they are the same for all classes of reserves for a particular mineral commodity market. Thus, the total Hotelling rent shown in Figure 3-2 is simply the Hotelling rent earned on marginal reserves (C_mP) times total reserves (0M).

Those reserves whose marginal extraction costs are below those of the marginal reserves in class M are called inframarginal reserves. As a result of their relatively low costs, they yield additional profits when they are exploited. Mineral economists refer to these additional profits as *Ricardian rents*. In Figure 3-2, the Ricardian rents per unit of output equal C_1C_m for reserves in class A, C_2C_m for reserves in class B, and so on.

Unless technical or other considerations intervene, mineral producers will generally exploit first those reserves that have relatively low production costs and thus high Ricardian rents (like classes A and B). This implies that the reserves currently being extracted have lower costs than the average of all reserves and that their Ricardian rents are likely to be above average.

Since reserves by definition are known and profitable to exploit, they are assets in the sense that they have value in the marketplace. Although mineral resources other than those classified as reserves might have incompletely defined characteristics (in terms of costs and quantities) or be currently unprofitable to exploit, they still may command a positive price in the marketplace. Petroleum companies, for example, pay millions of dollars for offshore leases to explore for oil deposits that are not yet proved reserves. Mining companies pay for and retain subeconomic deposits. The option of developing such deposits in the future has a positive value because the price may rise, or some other developments may make the deposits economic.

Thus, a full accounting of subsoil assets should consider not only reserves, but also other mineral resources with positive market value. In the case of reserves, market value may reflect Hotelling rent, Ricardian rent, and option value.[4] In the case of mineral resources other than reserves, a positive market value is due solely to their option value.

Key Definitions in Mineral Accounting

Changes in the value of the mineral stock come about through additions, depletions, and revaluations of reserves.

- *Additions* are the increases in the value of reserves over time due to reserve augmentations. They are calculated as the sum of the price of new reserves times the quantity of new reserves for each reserve class.
- *Depletions* are the decreases in the value of reserves over time due to extraction. They are similar to capital consumption (depreciation) and parallel the concept of additions.
- *Revaluations* are changes in the value of reserves due to price changes. They measure the residual change in the value of reserves after correcting for additions and depletions.

Techniques for Valuing Mineral Assets

As noted in the last section, the major challenge in extending the national accounts to include subsoil minerals is to broaden the treatment of mineral assets to include additions and depletions and to incorporate depletion in the production accounts. This task involves estimating the value of the subsoil assets. A specific subsoil asset consists of a quantity

---

[4]The total value of reserves is $V = \sum_i v_i R_i$, where v_i is the unit value of reserves in class i $(i = A, B, \ldots, M)$, and R_i is the quantity of reserves of class i.

SUBSOIL ASSET

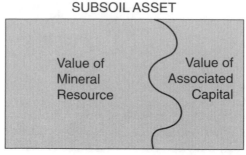

FIGURE 3-3   The Two Components of Subsoil Assets.

of a mineral resource and the invested capital associated with finding and developing that resource. Invested capital includes physical structures such as roads and shafts, as well as capitalized exploration and drilling expenses. The total value of the subsoil assets equals the sum of the value of the mineral and the value of the associated capital (see Figure 3-3). Currently, U.S. national economic accounts include the value of the associated capital, but exclude the value of the mineral resource. One of the goals of natural-resource accounting is to estimate the total value of subsoil assets and to separate this estimate into the value of the mineral and the value of the associated capital. An additional goal is to track over time changes in the value of the stock that result from additions, depletions, and revaluations.

Three alternative methodologies are used in valuing mineral resources: (1) transaction prices, (2) replacement value, and (3) net present value. In developing its mineral accounts, BEA used one version of the first method and four versions of the third. This section explains the basic elements of each approach.

Transaction Prices

The most straightforward approach to valuing mineral resources relies on market transaction prices. This is the standard approach used across the national economic accounts for capital assets. When resources of petroleum, copper, gold, and other minerals are sold, the value of the transaction provides a basis for calculating the market value of the mineral component of the asset.

A close look at the transaction-prices approach reveals, however, a number of difficulties that need to be resolved. The major difficulty is that a market transaction usually encompasses a number of assets and liabilities, such as the associated capital (e.g., surface roads, shafts, and refining operations), taxes, royalty obligations, and environmental liabilities. Because the transaction usually includes not only the mineral resources, but also associated capital, the value of the capital must be subtracted to obtain the mineral value. In addition, the property is usually encumbered with royalty obligations to prior owners or to owners of the land. Many mineral properties also have associated environmental problems, such as contaminated soils and water, and they may even be involved in complicated legal disputes, such as connection to a Superfund site with joint and several liability. Some of these associated assets and liabilities (such as mining structures) are true social costs or assets, while others (such as royalty obligations) are factor payments.

Another difficulty with using transaction prices is the sporadic nature of the transactions. The infrequency of the transactions, coupled with the heterogeneity of the grade of the resource, makes it difficult to apply the transaction price for one grade or location of the resource to other grades in other locations.

Because of the complex assortment of assets and liabilities associated with transactions of mineral resources, the price must be adjusted to obtain the value of a resource. As noted above, the working capital and the associated capital must be subtracted from the transaction price, while any extrinsic environmental liabilities should be added, as should any factor payments, such as royalties or taxes, to obtain the value of the underlying resource.

Box 3-1 provides an example of how to adjust the transaction price to obtain the market value of a mineral resource for a hypothetical sale involving the purchase of 500,000 barrels of oil. In this example, the buyer pays $2 million for a property containing 500,000 barrels of oil, and this is recorded as the transaction value. Attached to those reserves is a long-term debt of $1.0 million; this liability must be added to the purchase price. If the acquired reserves also include associated working capital of $0.2 million, this amount must be deducted from the purchase price. Correcting for these two items creates an effective purchase price or market value of the asset of $2.8 million.

An additional issue arises because of payments such as future taxes and royalties. In acquiring the above property, the new owner must, for example, pay a 10 percent overriding royalty to the landowner. Such payments should be included in the value of the resource even though they do not accrue to the seller of the property. In the example shown in

---

Box 3-1
Transaction Price Method[a]

Recorded Dollar Transaction (500,000 barrels) .... $2.0 million
    Adjustments
        Add: assumed liabilities .... $1.0 million
        Subtract: working capital .... $0.2 million

Effective Purchase Price of Asset .... $2.8 million
        Add: present value of taxes, royalty transfers ... $0.6 million

Value of Assets .... $3.4 million
        Subtract: value of associated capital .... $0.8 million

Value of Petroleum Reserve .... $2.6 million

_____

[a]This methodology is not followed in the conventional accounts. For instance, in valuing the stock of cars, we do not subtract tax credits, nor do we add in future liabilities such as property taxes. Similarly, to the extent that royalties are regarded as a sharing of profits (like dividends), they should not affect the value of an asset; to the extent that royalties are actually a deferred part of the purchase price, they can be capitalized to increase the value of an asset.

Box 3-1, future royalties and taxes are assumed to have a present value of $0.6 million. These payments introduce a major new complication because taxes and royalties depend on future production. Not only are they uncertain, but they also cannot be easily estimated from market or transaction data. One approach is to adjust the transaction price by marking up the value of the transaction by a certain amount. Adelman and Watkins (1996:4), for example, suggest that 27 percent be added to the "effective purchase price" to account for transfers. After adjusting for royalties, this yields a social asset value for the above property of $3.4 million. The final adjustment is for associated capital, which is assumed to have a value of $0.8 million. After this amount is subtracted, the estimated social value of the underlying petroleum reserve is calculated to be $2.6 million.

Replacement Value

A second approach uses the costs of replacing mineral assets to determine their value. Under this approach, it is assumed that firms have an incentive to undertake investments to find new resources up to the point where the additional cost of finding one more unit just equals the price at which firms can buy that unit—that is, up to the market value. Therefore, the additional or marginal cost of finding a mineral resource should be close to its market price. Associated with this approach, however, are many of the same issues discussed above under transaction prices. For example, a particular replacement cost is relevant only for valuing deposits of comparable quality and cannot be used to value resources of another grade. This point can be illustrated using Figure 3-2. Assume that exploration is resulting in the discovery of resources of class M. The market value of this class would be a function of the difference between 0P and production cost $0C_M$. It would be profitable for firms to continue exploring for such deposits until the finding costs (that is, the replacement costs) just reached the value of this class of resource. However, the replacement cost of class M cannot be used to value other classes, such as class A, which have a lower extraction cost and therefore a higher value. Because of cost differences, using class M to value classes A through L would yield an underestimate of the value of these reserves.

Net Present Value

A third valuation technique, the net present value or NPV method, entails forecasting the stream of future net revenues a mineral resource would generate if exploited optimally, and then discounting this revenue stream using an appropriate cost of capital.[5] Under certain conditions— such as no taxes—the sum of the discounted revenue values from each time period will equal the market value of the resource. For example, assume that a 100 million-ounce gold asset generates a stream of net revenues (after accounting for all extraction and processing costs) that, when discounted at a rate of 10 percent per year, has a present value of $1.5 billion. According to this approach, the value of the asset is taken to be $1.5 billion. If the value of the plant, equipment, and other invested capital ultimately associated with the asset is estimated to be $500 million, the current value of the gold reserves is $1 billion, and their unit value is $10 per ounce. Again, as with the previous two methods, each class of resource should be separately valued, since the stream of revenues from a higher class of resource will be greater than that from a lower class.

A special case of the NPV approach, known as the Hotelling valua-

---

[5]The appropriate discount rate for energy and environmental resources is debatable. See Lind (1990, 1997), Schelling (1995), and Portney and Weyant (1999).

tion principle (see Miller and Upton, 1985), avoids the difficulties of fore-casting future net revenues and then discounting them back to the present. This approach makes the strong and generally unrealistic assumption that the unit value of a resource grows at exactly the same rate as the appropriate discount rate. In the above example, this would imply that the unit value of the gold resource would grow at the discount rate of 10 percent per year; that is, the unit value would be $10 in the first year, $11 in the next year, $12.1 in the following year, and so forth. Under this assumption, the present value of the resource would easily be calculated as the current period's resource price multiplied by the current physical stock of the resource. Under a further set of assumptions, such as homo-geneous resources and constant extraction costs, the current period re-source price is simply the current net revenue (unit price less unit extrac-tion cost).

For example, assume that in a given year the United States has 100 million ounces of homogeneous gold reserves, that the price of gold in that year is $350 per ounce, and that the average extraction cost is $335 per ounce. Under the Hotelling valuation principle, the price of the gold reserves would be $15 per ounce, and the total value of the gold assets would be calculated as $1.5 billion. Note that it would still be necessary to deduct the value of capital from the $1.5 billion to obtain the value of the mineral reserve. Again, for this approach to be valid, the per unit price of gold reserves ($15 in this example) would need to grow at the discount rate appropriate for these assets.

BEA'S VALUATION OF SUBSOIL MINERALS

This section presents a more detailed description of BEA's valuation methods (as set forth in Bureau of Economic Analysis, 1994b). In the absence of observable market prices for reserves, BEA estimates mineral reserve and flow values using five valuation methods. These calculations are performed for reserves of fuel minerals (petroleum, natural gas, and coal) and other minerals (uranium, iron ore, copper, lead, zinc, gold, sil-ver, molybdenum, phosphate rock, sulfur, boron, diatomite, gypsum, and potash) for each year from 1958 through 1991 (oil and gas figures are calculated from 1947 to 1991). In addition, aggregate stock and flow values for five mineral categories (oil, gas, coal, metals, and other miner-als) are entered in the appropriate rows and columns of the IEESA Asset Account for 1987. This section first examines the five methods used by BEA in estimating mineral values, along with the data they require, and then describes BEA's findings. Box 3-2 provides definitions of the sym-bols used in minerals accounting.

Box 3-2
Definitions of Symbols and Basic Concepts in Minerals Accounting

For this discussion, assume that there is only one class of a mineral reserve, that extraction costs are constant, and that the unit value of the reserve rises at the social rate of discount. Variables are:

R_t = total quantity of reserves of the mineral commodity at year end
H_t = unit value of the reserves (say, petroleum reserves), which equals Hotelling rent under the above assumptions
A_t = quantity of new reserves discovered during the year
q_t = quantity of extraction or production during the year
V_t = total value of the reserves at year end

In a given year, petroleum firms might discover new reserves totaling A_t. Then the additions are given by:

$$\text{additions}_t = H_t A_t \tag{3.1}$$

During that year, petroleum production, and therefore depletion of existing reserves, is measured by q_t. Depletion is, under the special assumptions listed above, quantity times the value of reserves:

$$\text{depletions}_t = H_t q_t \tag{3.2}$$

The total value of reserves at year end is:

$$\text{value of reserves} = V_t = H_t R_t \tag{3.3}$$

The change in the value from the end of year $t-1$ to the end of year t is given by:

$$\text{change in value of reserves} = V_t - V_{t-1} = H_t R_t - H_{t-1} R_{t-1} \tag{3.4}$$

Revaluations are the change in the value corrected for the value of additions and depletions:

$$\text{revaluation} = H_t R_t - H_{t-1} R_{t-1} - H_t A_t + H_t q_t \tag{3.5}$$

BEA's Five Basic Valuation Methods

Current Rent Method I

Current rent methods I and II are NPV methods based on the Hotelling valuation principle. The attraction of the Hotelling valuation principle is the ease with which the calculation can be performed, avoiding the need to forecast mineral prices and to assume an explicit discount factor. In both methods, the value of the aggregate stock is calculated as the net price times the quantity of reserves, where the net price is as described below. Additions or depletions are similarly calculated as net price times the quantity of additions or depletions. One of the difficulties with this approach is that the Hotelling valuation principle tends to provide a systematic overvaluation of reserves, the reason for which is discussed in a later section.

Current rent methods I and II are quite similar in construction. They differ primarily in the method of adjusting for the value of associated capital. (The algebra of the different formulas is shown in the boxes in this section.) Current rent method I (see Box 3-3) uses the normal rate of return on capital to determine the return on associated capital in the mining industry that should be subtracted from revenues. It then calcu-

---

Box 3-3
Formulas for Current Rent Method I

total mineral reserve value$_t$ = V_t = $[p_t - a_t] R_t - rR_tK_t/q_t - R_tD_t/q_t$

$$= [p_t - a_t - rK_t/q_t - D_t/q_t] \times R_t$$

additions$_t$ = $[p_t - a_t - rK_t/q_t - D_t/q_t] \times A_t$

depletions$_t$ = $[p_t - a_t - rK_t/q_t - D_t/q_t] \times q_t$

revaluations$_t$ = $V_t - V_{t-1}$ + depletions$_t$ − additions$_t$

where

   V_t = value of mineral reserves
   p_t = price of commodity
   a_t = average cost of current production
   R_t = total quantity of reserves
   r = average rate of return on capital
   K_t = value of associated capital, valued at current replacement cost
   q_t = total quantity extracted
   D_t = depreciation of associated capital
   A_t = quantity of discoveries of new reserves
   additions$_t$ = value of discoveries of new reserves
   depletions$_t$ = value of depletions
   revaluations$_t$ = change in value of reserves corrected for depletions and additions

The revaluation term is not directly calculated; it will include any errors in calculating additions, depletions, and opening and closing stock values.

Box 3-4
Formulas for Current Rent Method II

total mineral reserve value$_t$ = V_t = $[p_t - a_t - K_t/R_t]\, R_t$

additions$_t$ = $[p_t - a_t - K_t/R_t] \times A_t$

depletions$_t$ = $[p_t - a_t - K_t/R_t] \times q_t$

revaluations$_t$ = $V_t - V_{t-1}$ + depletions$_t$ − additions$_t$

where variables are as defined in Box 3-3.

---

lates the "resource rent per unit of reserve" by taking the net profits from mining, subtracting the return and depreciation on the associated capital, and dividing that sum (called "resource rent" by BEA) by the quantity of resource extracted during the year. The method thus yields an estimate of the unit value of the reserves currently extracted.

To calculate the total value of the mineral reserve, the current resource rent per unit is multiplied by the total reserves, in the spirit of the Hotelling valuation principle. Additions and depletions are calculated as those quantities times the resource rent per unit. Revaluations are simply the residual of the change in the value of the stocks plus depletions minus additions. It has been observed that the value of the stock can be highly volatile; this volatility is due primarily to the revaluation effect.

Current Rent Method II

Current rent method II is virtually identical to current rent method I. The only difference is in the method of adjusting for associated capital. The value of the associated capital is subtracted from the total value of the mineral asset to obtain mineral-reserve values in current rent method II. Again employing the Hotelling valuation approach, the total value of the mineral asset (including the value of the associated capital) is calculated as the per unit net revenue times the total quantity of reserves. The total value of the mineral reserve is then calculated as the total value of the asset value minus the value of the associated capital. The unit resource value, which is used to price additions and depletions, is just this total reserve value divided by the total quantity of reserves. This approach is defined algebraically in Box 3-4.

As is discussed below, both current rent methods have major advantages in that they are easy to calculate on the basis of data BEA currently uses in its accounts (primarily profits and capital stock and consumption data). They both suffer from the serious disadvantage that they rely on the Hotelling valuation principle, thereby tending to overvalue reserves.

Net Present Value Estimates

If the basic assumptions of the Hotelling valuation principle do not hold—and there is strong evidence that they do not, as discussed below—life becomes much more complicated for national accountants. One approach that is sound from an economic point of view is to value reserves by estimating the present discounted value of net revenues. To render the present value approach workable, BEA makes three simplifying assumptions. First, it assumes that the quantity of extractions from an addition to proved reserves is the same in each year of a field's life. The quantity of depletions in any year is assumed to result equally from all vintages (cohorts) still in the stock, i.e., all vintages whose current age is less than the assumed life. Second, the life for a new addition is assumed to be 16 years until 1972 and 12 years thereafter. Third, BEA assumes that the discount rate applied to future revenues is constant at a rate of either 3 percent per year or 10 percent per year above the rate of growth of the net revenues (where the latter equals the rate of growth of the price of the resource).[6]

These assumptions lead to a tractable set of calculations. The present discounted value of the mineral stock as calculated using this present value method is simply the stock and flow values calculated with current rent method II, multiplied by a "discount factor" of between 0.86 and 0.89 for the 3 percent discount rate and between 0.63 and 0.70 for the 10 percent discount rate.[7] The calculated values are, then, lower than the values derived using current rent method II, with the difference depending on the discount rate employed.

Additions and depletions are then calculated in a manner similar to that used with current rent method II. The average unit reserve value is

---

[6]According to BEA, the rates were chosen to illustrate the effects of a broad range of approaches. The 3 percent per year discount rate has been used by some researchers to approximate the rate of time preference, while the 10 percent rate has been used by some researchers to approximate the long-term real rate of return to business investment.

[7]At the 3 percent discount rate, the 0.86 discount factor holds for the years 1958 through 1977, with the rate edging upward thereafter as a result of commingling of reserves that were developed prior to 1973 (which BEA assumes are extracted over 16 years) with those developed in 1973 or later (for which a 12-year life is assumed). For the 10 percent discount rate, the 0.63 factor holds for the years 1958 through 1974. In 1987, the year for which BEA calculates a more complete set of satellite accounts, the rate is 0.88 for the 3 percent discount rate and 0.69 for the 10 percent discount rate.

Box 3-5
Formulas for Net Present Value Method

total mineral reserve value$_t$ @ 3 percent discount rate = 0.88 $[p_t - a_t] R_t - 0.88 K_t$
total mineral reserve value$_t$ @ 10 percent discount rate = 0.69 $[p_t - a_t] R_t - 0.69 K_t$

additions$_t$ @ 3 percent discount rate = 0.84 $[p_t - a_t - K_t/R_t] \times A_t$
additions$_t$ @ 10 percent discount rate = 0.59 $[p_t - a_t - K_t/R_t] \times A_t$

depletions$_t$ @ 3 percent discount rate = 0.83 $[p_t - a_t - K_t/R_t] \times q_t$
depletions$_t$ @ 10 percent discount rate = 0.60 $[p_t - a_t - K_t/R_t] \times q_t$

revaluations$_t$ = $V_t - V_{t-1}$ + depletions$_t$ – additions$_t$

where variables are as defined in Box 3-3.

Note: The numerical values in this box apply to 1987. As explained in the text, slightly different values will apply for different years.

---

calculated by dividing the total reserve value by the quantity of reserves, and then using this unit value to value additions and depletions. Additions would be calculated as 84 percent of the value of additions according to current rent method II if the discount rate is 3 percent per year, and 59 percent of the value of additions according to current rent method II if the discount rate is 10 percent. The calculated value of depletions would be 83 percent of the value of depletions under current rent method II at a 3 percent discount rate, and 60 percent at a 10 percent discount rate.

In summary, the present value method as implemented by BEA takes the values of additions, depletions, and stocks calculated according to current rent method II and multiplies them by discount factors of between 59 and 88 percent. The reason for the discount is straightforward. Under current rent method II, which relies on the Hotelling valuation principle, it is assumed that net revenues rise at the discount rate. Under the present value approach, net revenues are assumed to rise at rates that are 3 or 10 percent slower than the discount rate applicable to mineral assets. The higher percentage is the discrepancy between the rise in net revenues and the discount rate; the lower is the discount factor. The NPV approach is shown in Box 3-5.[8]

---

[8]As with the calculation of mineral values, the factors shown in Box 3-5 vary depending on the year of the analysis. The factors reported are those for the 1987 calculation. The factors differ in the various formulas because of the differing treatment of the timing of depletions and additions from reserves.

Replacement Cost

The fourth method of calculating the value of the mineral stock is used only for oil and gas reserves. Despite its name, this approach is similar to the NPV method, not to the replacement cost method described earlier. It adopts the approach of Adelman (1990), who calculates the present value of an oil field using special assumptions. It is assumed that the production from an oil or gas field declines exponentially over time. Under the assumption that the decline rate is constant and that the net revenue rises at a fixed constant rate that is less than the discount rate, a barrel factor is calculated. This barrel factor is multiplied times net revenue to obtain the present value of the reserves. Adelman estimates that the barrel factor is usually around 0.5. BEA does not give the barrel factor used in its calculations, which should vary by deposit and depend on the rate at which future cash flows are discounted, but we estimate that it averages approximately 0.375.

The value of the asset—calculated with current rent method II using the Hotelling valuation principle—is then multiplied by the barrel factor. The justification is that this NPV approach, unlike the Hotelling approach, takes the physical specifics of oil and gas extraction into account and accordingly adjusts the unit value of reserves downward. As with the NPV approach discussed in the last section, this adjustment accounts for the overvaluation inherent in the Hotelling valuation principle.

Once the value has been adjusted downward, BEA must again subtract the value of capital associated with the asset. With this method, the value of capital associated with each unit of existing reserves is assumed to be the current-year expenditure on exploration and development for oil and gas, divided by the quantity of oil and gas extracted during the year. This approach is loosely based on Adelman's suggestion that the value of capital associated with a unit of production can be approximated by measuring the value of capital associated with finding new reserves. The replacement cost method is shown in Box 3-6.

Transaction Price Method

When oil and gas firms desire additional reserves, they can either buy them from other firms or find new ones through exploration and development. In the absence of risk, taxes, and other complications, the transaction price of purchasing new reserves should represent the market value of those reserves. For this reason, according to BEA, "if available, transaction prices are ideal for valuing reserves" (Bureau of Economic Analysis, 1994b:57).

In fact, transactions in reserves are few and far between outside of oil and gas, and even in oil and gas suffer from problems discussed above.

Box 3-6
Formulas for Replacement Cost Method

total mineral reserve value$_t$ = V_t = {$0.375[p_t - a_t] - Z_t/q_t$}$R_t$

additions$_t$ = {$0.375 [p_t - a_t] - Z_t/q_t$} x A$_t$

depletions$_t$ = {$0.375 [p_t - a_t] - Z_t/q_t$} x q$_t$

revaluations$_t$ = $V_t - V_{t-1}$ + depletions$_t$ - additions$_t$

where Z_t = value of exploration and development expenditures in year t, and other variables are as defined in Box 3-3.

To estimate transaction prices, BEA derived prices from publicly available data on the activities of large energy-producing firms for the period 1977 to 1991. The gross value of reserves was estimated by dividing expenditures for the purchase of the rights to the proved reserves by the quantity of purchased reserves. The result was then adjusted for associated capital using the same method as in current rent method II. The transaction price method is shown in Box 3-7.

Data Requirements

On the whole, the five valuation methods used by BEA are relatively parsimonious, and therefore the data requirements are not unduly burdensome. For quantity data, only reserves are considered, so the quantities of mineral stocks are easy to obtain. Most of the data required for valuation under the five methods either are already used by BEA in their

---

Box 3-7
Formulas for Transaction Price Method

total mineral reserve value$_t$ = V_t = $(TV_t/TQ_t - K_t/R_t) R_t$

additions$_t$ = $(TV_t/TQ_t - K_t/R_t)$ x A$_t$

depletions$_t$ = $(TV_t/TQ_t - K_t/R_t)$ x q$_t$

revaluations$_t$ = $V_t - V_{t-1}$ + depletions$_t$ - additions$_t$

where TV_t = value of reserve transactions, and TQ_t = total quantity of reserves transacted, and other variables are as defined in Box 3-3.

---

construction of the NIPA or are publicly available or available at a modest cost from private sources. Constructing the accounts for subsoil minerals, therefore, required no independent data collection or survey by BEA. Nevertheless, there is no single consolidated source for the data needed, and considerable effort was expended by BEA staff in collecting the data.

Preliminary Results

The first set of estimates in the IEESA contains many important and useful conclusions. We highlight some of the key findings in this section.[9]

The calculations present a number of interesting findings for the overall economy. All five evaluation methods indicate that the value of the stock of oil and gas reserves in the United States exceeds the value for all other minerals combined. For all subsoil minerals, the calculated value of reserve additions has approximately equaled the value of depletions over the 1957-1991 period. Consequently, the value of reserves (in constant prices) has changed little during the reporting period. BEA finds that the value of the mineral component of a mineral asset is about 2 to 4 times the value of the associated capital, so the value of the mineral makes up 67 to 80 percent of the total value of any mineral asset.

The results are also helpful in understanding returns to capital of U.S. companies. Standard rate-of-return measures include profits on mineral assets in the numerator, but exclude the value of mineral reserves in the denominator. Gross rates of return for all private capital decline from 16 percent per year if mineral reserves are excluded to 14-15 percent if mineral reserves are included. BEA does not present net returns, however. Because net post-tax returns on nonfinancial corporate capital have averaged around 6 percent per year over the last three decades, our estimate of the profitability of American corporations would be significantly modified if the 1-2 percentage point decline in the gross return carried over to the net return.

In quantity terms, the physical stock of aggregate metal reserves has tended to decline over time, while the physical stock of coal reserves has increased. Quantities of oil, gas, and industrial minerals ("other minerals" in BEA's five broad categories) have remained stable. Revaluations have tended to be positive primarily because the prices of most subsoil minerals have risen over the period under investigation.

---

[9]These findings are presented in Bureau of Economic Analysis (1994b) and summarized in Table 4-1 in Chapter 4 of this report.

BEA estimates the value of the nation's stock of mineral reserves, after deduction of associated capital, to be between $471 billion (current rent method I) and $916 billion (current rent method II) for 1991; this figure amounts to between 3 and 7 percent of the value of produced assets (existing produced structures, equipment, and inventories). Current rent method II yields the highest stock and flow values for all mineral types. Current rent method I yields the lowest values for coal, metals, and other minerals, while the transaction price method yields the lowest value for oil, and the replacement cost method yields the lowest value for gas. (Recall that these last two methods are used only for oil and gas.) Given the algebra of the different valuation techniques, it is not surprising that the replacement cost method yields lower values than the current rent methods for gas since the replacement cost method is really current rent method II multiplied by 0.375.

One important question concerns the impact of including subsoil minerals in the overall national accounts. In 1987, the year for which BEA presents the IEESA asset accounts, the calculated value of reserve additions roughly offsets reserve depletions, so including mineral assets in the NIPA for that year would not substantially alter the estimate of the level of net domestic product (NDP). It would, however, increase the level of GDP by between $17 and $65 billion (0.4 to 1.4 percent of GDP), depending on the method used to value reserve additions. The only year in which the mineral accounts would have a substantial impact on the growth of real GDP or NDP is 1970, the year Alaskan reserves were added. Box 3-8 shows the calculations of real GDP (in 1987 prices) with and without mineral additions for that year. The large surge of oil reserves erases the recession of 1970 and leads to a downturn in growth in 1971. While this kind of volatility is unique in the period analyzed by BEA, it does indicate that introducing minerals into the accounts might lead to large changes in measured output that would reflect primarily changes in mineral reserves.

EVALUATION OF BEA'S APPROACH

This section evaluates the methodology of BEA's preliminary approach to accounting for subsoil minerals. We begin with the advantages of the approach and then review some issues and concerns.

Advantages

Feasibility

Phase I of BEA's plan for extending the national accounts to include supplemental mineral accounts is now complete. In accordance with the

Box 3-8
Growth in Real Gross Domestic Product and Net Domestic Product With and Without Mineral Additions[a]

	(1) Conventional GDP	(2) GDP with Mineral Additions
1969	2.72	2.37
1970	0.03	3.14
1971	2.85	− 0.08

	(3) Conventional NDP	(4) NDP with Mineral Additions and Depletions
1969	2.53	2.13
1970	− 0.40	2.98
1971	2.71	− 0.48

[a]Percent per year.
Source: Conventional GDP and NDP in 1987 prices were calculated by BEA (*U.S. Congress, Economic Report of the President, 1995*). GDP with mineral additions was calculated based on data in columns (1) and (3) and estimates of mineral additions and depletions from Bureau of Economic Analysis (1994b:60). Mineral additions and depletions in this calculation rely on current rent method I.

recommendations of the United Nations System of National Accounts (SNA), BEA limited the focus of Phase I to mineral reserves. This is probably the simplest of the natural-resource sectors to include because the output is completely contained in the current national accounts and involves primarily estimating and valuing reserve changes. The data, although obtained from various sources, are publicly available from the (former) Bureau of Mines, the U.S. Geological Survey, the U.S. Department of Energy, and the Bureau of the Census. Some minor adjustments of the data were needed in cases where the definition of reserves changed over time.

BEA began this work in 1992 and completed it in April 1994. Given the late start and limited resources of the U.S. natural-resource accounting effort, along with the sparsity of observable market prices with which to value mineral additions, depletions, and stocks, the progress made by BEA to date is remarkable. Furthermore, the task was completed by a group of eight BEA officials working part time on this assignment while continuing with their regular duties. The result is a partially completed satellite account that fits into the current definitions of the U.S. NIPA and can be readily prepared in a short amount of time. BEA's approach is therefore clearly feasible and relatively inexpensive.

Consistency with Other Valuation and Accounting Frameworks

BEA treats mineral additions in parallel with other forms of capital formation. In this respect, the U.S. accounts differ from the System of Integrated Environmental and Economic Accounting (SEEA), an alternative satellite accounting system proposed by the United Nations. In both accounting systems, depletions are treated as depreciations of the fixed capital stock. Under the SEEA, however, additions are not included as income and do not appear in the production accounts as capital formation.

In calculating GDP, the SEEA considers as capital formation only investments in "made capital" and not mineral finds, treating discoveries as an "off-book" entry. This approach avoids the volatility associated with mineral finds, which, if included in GDP, makes GDP a volatile series (see Box 3-8). BEA, on the other hand, treats mineral assets on the same basis as fixed capital. For example, according to BEA calculations, booking the exceptional Alaskan oil finds in 1970 augmented the existing stock of U.S. oil assets by nearly 50 percent, or almost $100 billion in 1987 prices, despite exploration investments on these reserves that were only a fraction of this amount. Including the increase in mineral reserves in private investment would have increased gross investment by 26 percent in 1970 and would have increased net investment by 42 percent. As is seen in Box 3-8, the trend in real nonminerals GDP growth would have been seriously distorted, wiping out the 1970 recession and causing an apparent recession in 1971. Thus, while including mineral additions as capital formation treats made and natural capital augmentations in a parallel fashion, the aggregate GDP series may become more volatile and may not accurately reflect movements in production and employment.

A second concern with treating mineral additions as capital formation is that the two do not necessarily have the same effect on the economy. In particular, when fixed capital is added to the capital stock, payments have been made to the factors of production involved in producing the capital. Mineral-stock additions, in contrast, reveal themselves as increases in land value, which are balance sheet adjustments rather than payments to factors of production. It is for this reason that the United Nations SEEA approach omits additions from net investment in the production accounts and introduces a reconciliation term in the asset accounts to capture additions.

Finally, it has been argued by some that mineral stocks are inventory and should be treated as such in the NIPA. BEA chooses to treat mineral stocks as fixed capital, suggesting that, just as with produced fixed capital, expenditures of materials and labor are needed to produce these mineral assets, which in turn yield a stream of output over an extended period of time. The treatment of mineral stocks then becomes consistent with the treatment of traditional capital in the NIPA. Of course, the concept of a satellite account allows individual policy researchers to take the information in these accounts and make their own adjustments to the NIPA. The BEA approach is just one potential way of treating natural capital formation and depletion.

In terms of valuation methodology, the BEA approach is consistent with current mineral asset valuation theory.

Utility

BEA presents an IEESA Asset Account and an IEESA Product Account that supplement the NIPA. Researchers, businesses, and policy makers can use the satellite accounts to adjust output and income measures as they see fit, focusing on any or all of the five valuation methods used by BEA. Moreover, BEA presents separate entries for five types of mineral assets, including three types of fuels, and an aggregate mineral category.

This level of detail makes the satellite accounts useful to policy makers who wish to focus on particular mineral issues. The data on the value of mineral stocks, additions, depletions, and revaluations (the residual) are given annually for the 1947-1991 period for oil and gas (the two most important mineral groupings in terms of total stock value) and from 1958 to 1991 for the other three mineral groupings. The constant (1987) dollar figures for the aggregate mineral stock show a price-weighted index of the stock, as well as of additions and depletions to the aggregate, and are useful for determining whether the aggregate price-weighted quantity of U.S. mineral reserves is changing over time. One of the important findings from the BEA data is that the index of the total constant-price stock of mineral assets has been approximately constant from 1957 to 1991. This implies that the nation has on average replaced reserve depletions with an equivalent quantity of reserve additions (or, more precisely, quantities of reserve additions and depletions of different minerals weighted by 1987 prices).

Issues and Concerns

BEA's approach to calculating mineral stock and flow values raises a number of issues related both to measurement problems and to conceptual concerns with the individual valuation techniques. Some of these issues are intrinsic to any accounting approach in which data on prices or quantities must be imputed or constructed, while other issues arise for particular methodologies. The major issues are reviewed here.

Heterogeneity of Reserves

A major problem with most accounting approaches is that they assume all reserves are homogeneous in terms of grade and costs. For example, under the Hotelling valuation principle, average extraction cost should be calculated as the average cost of extraction from all reserve classes. In practice, most techniques use the extraction cost of currently extracted reserves. The reality is that a nation's reserves are not all in one cost class. It has already been noted that reserves are likely to exist in a number of classes, ranging from high quality (low cost) to low quality (high cost). Resource accounting, such as that in the current IEESA, generally treats the entire national stock as one heterogeneous deposit whose value is calculated by multiplying the average unit value of that reserve by the quantity of the reserve.

An example will illustrate the issues raised by resource heterogeneity. Suppose that a nation owns 100 million ounces of subsoil gold reserves whose total value is $1 billion, for an average unit value of $10 per ounce. In a given year, the nation extracts 1 million ounces, with no additions, and the value of the remaining reserves with unchanging gold prices is $989 million. Accordingly, the depletion is measured at $11 million, with an average value of $11 per ounce extracted. This pattern is typical of many extraction profiles in which the lowest-cost and highest-value resources are extracted first.

Note that the correct depletion charge is the value of the extracted ore times the quantity extracted, for a total of $11 million. If we were instead to use the average value of the ore of $10 per ounce to value depletion, we would be underestimating depletion at $10 million rather than $11 million. Moreover, if we used the value of the extracted reserve to value the remaining reserves of 99 million ounces, we would incorrectly value reserves at 99 x $11 = $1089 million, rather than the correct $989 million. This example shows that with reserve heterogeneity, using the average reserve value to estimate depletion is likely to understate depletion, while using the value of the extracted resource to value remaining reserves is likely to overstate the value of reserves.

This example is useful because common practice in constructing national resource accounts, and one of BEA's approaches, uses the average value of the extracted resource to value the entire reserve stock. Nor can average costs from current production be used to calculate the net present value of additions. Because of the random quality of additions, it is not possible to determine whether additions will be undervalued or overvalued using these cost data. Heterogeneity of reserves poses problems for the transactions approach because transaction values need not reflect the average value of the total reserves, as those parcels of reserves sold in any one period may have a quality above or below the average. All these problems of heterogeneity are particularly severe for metals, because there is a clear tendency for ore grades to fall over time. The issue is less clear for petroleum because new findings may have lower cost than current production, but the general trend in petroleum has been for lower finding rates per unit drilling.

Putting the point differently, the difficulty in valuing the stocks and flows arises because the prices of reserves are not readily available. Although the commodities, such as gold and oil, trade frequently, the underlying assets tend to trade infrequently. There is no organized market for oil or gold properties, and there is such great heterogeneity in these assets that there is no standard for classifying them as there is for oil or gold (in terms of sulfur content, purity, and the like). When reserves are transacted, the prices are not generally publicly available, which means the reserve prices are generally not observable. A further difficulty is that the tendency is to observe the value of the total bundle of assets and liabilities (reserves, associated capital, environmental liabilities, royalty and tax obligations, and so on), so that even if the transaction price were observed, the price of the mineral reserve could not readily be determined. All these complications mean that the values of reserve stocks, additions, and depletions—which are essential for the construction of national accounts for subsoil assets by BEA and other statistical agencies—must be estimated using the relevant economic and financial theories of valuation.

In principle, the heterogeneity problem could be overcome by calculating reserve values for each reserve class and then aggregating across reserve classes. This approach is likely to be quite costly, and extraction data may not be available for all reserve classes, particularly those not yet being exploited. However, since these disaggregated calculations are not undertaken by BEA, its estimated values for the total reserve stock are likely to be too high for many of the minerals.

If in fact the lowest-cost and highest-value reserves are extracted first, the use of extraction costs from current depletion will provide a biased estimate of reserve values. All of the BEA valuation methods except the transaction cost method use an inappropriate measure of reserve values based on the cost of current extraction. Although BEA does not report total mineral asset and mineral resource values separately, the estimation bias in the asset value will flow through to the calculation of the mineral value that BEA does report in Table 1, rows 36 through 41 (Bureau of Economic Analysis, 1994a). The result will be an upward bias in the mineral-resource values calculated with current rent method II. Whether this bias carries through to the calculation of mineral-resource values in the other calculation methods is unknown since, as discussed below, the deductions for capital may be too high or too low with the other approaches.

A similar problem arises in valuing reserve additions, since BEA assumes they have the same characteristics as current depletions. Consequently, if the quantity of additions equals the quantity of depletions, the value of additions will equal the value of depletions, even though the grade of reserves may be quite different for depletions and additions. BEA's approach is likely to overvalue additions. With the best deposits extracted first, additions are likely to be of less value than current depletions. This discrepancy will affect the IEESA production account since with a lower value for additions, the adjusted GDP and NDP figures will be lower. The discrepancy also introduces a downward bias into the revaluations of minerals because of the overstatement of additions.

Measures of Resource Quantities

Although most of the issues in minerals accounting involve valuation, issues involving the quantity of reserves or resources are also important in a few areas.

The first of these issues relates to the comprehensiveness of the resource base considered by BEA. In constructing product and asset accounts, one is concerned with valuing the stock of the nation's mineral resources and estimating changes in the value of the stock due to depletions, additions, and revaluations. These quantities are measured with considerable uncertainty. An important issue here (as it is throughout the federal statistical system) is developing measures of accuracy, both for satellite accounts and the main accounts. Mineral resources other than reserves are often unknown or not well established and thus are also quite difficult to measure with any accuracy. In all cases, even where quantities are known, their value is not easily calculated. For example, resource class N in Figure 3-2 has an average current extraction cost above price; thus, according to the Hotelling valuation principle, its value is zero. All resources other than reserves (classes N and above in Figure 3-2) are assigned zero value. For both practical and economic reasons, BEA considers only reserves in its IEESA. Hence, BEA's asset account includes a blank row for measures of stocks and of additions to and depletions from unproved subsoil assets. Yet these nonreserve resources are likely to have some positive market value because of their option value.

A related flaw in the BEA preliminary accounting framework is that current additions to reserves produce no compensating depletion of non-reserve resources. Yet every ton of reserves comes from nonreserve resources. If nonreserve resources have economic value (as they certainly do in the case of many oil and gas properties), the result will be an upward bias in the current estimates of net capital formation (additions minus depletions) in mineral resources. The failure to consider nonreserve resources means that additions to, as well as depletions from, different categories of nonreserve mineral assets are ignored. For example, adjacent drilling may lead to moving a resource from the speculative to the hypothetical category or from an inferred submarginal resource to a demonstrated subeconomic resource (see Figure 3-1). Proven reserve quantities sometimes change dramatically because previously uncertain nonreserve resources are found to be economic (e.g., Alaskan oil). Because the option values of different grades will differ, the overall bias in mineral capital formation could be in either direction. The basic problem again is valuing nonreserve resources. BEA intends ultimately to include unproved resources as a part of non-produced environmental assets.

It is recognized that current estimates of mineral capital formation are incomplete and likely to be biased. BEA correctly notes that an operational methodology for valuing these nonreserve resources is not yet available. As with reserves, market prices based on resource transactions are not widely available, especially outside of oil and gas, and unit prices must be deduced using related economic series. Economists are currently involved in developing methods for valuing such resources. However, official natural-resource accounting procedures have without exception omitted nonreserve mineral assets. Fortunately, the omitted value may not be great.[10]

A final issue is that BEA values only a subset of U.S. mineral reserves. Omitted are several heavily mined industrial minerals such as sand and

[10]Kilburn (1990) suggests that the value of metalliferous ores in unexplored land is $Canadian 400 per 16.3 hectares. This equates to $US 7 per acre. Maintaining mineral claims in the United States requires an annual payment of $5 per acre, which, at a discount rate of 10 percent per year, equates to a net present value of $50 per acre. Hence, unexplored leased land with some indication of mineral potential would appear to have a market value of at least $50 acre. If 100 percent of the 387,000,000-acre U.S. land mass is mineable in the future (an obvious overestimate), the current value of subsoil mineral resources other than reserves is on the order of $19.4 billion at $50 per acre. Even when allowance is made for energy resources and industrial minerals and offshore petroleum potential, the total present value of resources, other than reserves, is unlikely to exceed $100 billion. BEA calculates a current reserve stock value of some $700 billion.

gravel, which may have small scarcity or Hotelling rents because of their superabundance but Ricardian rents because of their location. In production terms, BEA considers minerals that made up 77 percent of the value of mineral and energy production in the United States in 1970, a year in the middle of the available time series (Bureau of Mines, 1972). The BEA series is incomplete, but it values the most important mineral reserves, at least in terms of production value, in the United States.

Measurement of Associated Capital

Accounting for minerals poses serious issues of jointness of value of the mineral resource and the associated capital. Because these are complementary factors, dividing the total value between capital and minerals is difficult and involves somewhat arbitrary accounting conventions. Similarly, when minerals are extracted, the value of the existing mineral asset diminishes. Some of the decreased value is depreciation of capital, while some is depletion of the mineral reserve. The total depreciation in asset value due to extraction must be apportioned between the two in resource accounting. With capital depreciation being determined by guidelines that apply to capital more generally, the residual loss in value is then applied to depletion (see Cairns, 1997). The only rules that apply are that total depletions over the life of the asset must sum to the value of the resource, and the total depreciation over the life of the asset must sum to the value of installed capital. Hence in an accounting framework that must separate depletion from depreciation on an annual basis, the depletion numbers are based arbitrarily on the depreciation schedule chosen, being less than the total decrease in the value of the asset, but greater than zero. One comforting factor, however, is that although the breakdown in value or change in value between the capital component and the minerals component is somewhat arbitrary, this affects only the composition of the depletion and depreciation values and not the total asset value.

Once the value of a mineral asset has been calculated, the value of associated capital must be deducted to produce the mineral-reserve value. Only current rent method II and the transaction price method deduct associated capital appropriately. Because the value of the asset is likely to be overestimated through use of the Hotelling valuation principle, current rent method II will nevertheless tend to overvalue the stock of mineral reserves. Setting aside issues of heterogeneity and assuming that appropriate corrections are made for associated assets and liabilities, the transaction price method is the only method that in principle can provide unbiased estimates of the mineral value.

Current rent method I deducts depreciation and the gross return for capital per unit of extraction from gross price (see Box 3-3). Since one does not know whether this subtraction is more or less than the subtraction under current rent method II, one cannot say whether the calculated value of mineral-resource value using current rent method I will be too high or too low, even given its upward bias in the calculation of the total asset value due to use of the Hotelling valuation principle. In the case of the metals category, however, current rent method I gives negative values for the stock of metal reserves in the 1980s, which are clearly biased downward. It appears, then, that with current rent method I, the upward bias in measurement of total asset value due to use of the Hotelling valuation principle is outweighed by an excessive deduction for associated capital.

As noted in the previous section, the NPV method deducts some fraction of the value of associated capital. Doing so would make sense only if the value of the associated capital were thought to be less than its replacement cost. On average, one would expect the value of the associated capital to equal its replacement cost. The deduction for capital cost under the replacement cost method (see Box 3-6) also will generally not reflect the value of associated capital.

BEA includes exploration and finding costs as part of associated capital and then deducts these costs as part of the capital costs when valuing mineral reserves. This practice raises the question of what BEA is actually trying to value. If, for example, a gold deposit before the installation of any development expenditures or physical capital can be sold for $10 million dollars, some would suggest this is the value of the mineral reserves. BEA subtracts past exploration costs from this figure, and thus would value the mineral component of the property at less than $10 million. The former approach values the asset as a "gift of nature," while BEA values it as the product of previous human endeavor and charges the stock account with the cost of moving the mineral from the resource to the reserve category.

Early models of mineral value suggested that depletion can be calculated as current net revenue less capital depreciation less a return to capital, and BEA follows this approach with current rent method I. Subsequent research, however, has shown that this approach overestimates depletion (Cairns, 1997; Davis, 1997). As a result, estimates of depletion with current rent method I are too high, perhaps by as much as half. The depletion calculations with each of the other methods, including current rent method II, do not conform to any known depletion formulations, and the level or direction of measurement bias cannot be determined. Nevertheless, the panel's review indicates that the depletion calculations with current rent method I represent an *upper bound* on depletion. Moreover, according to Cairns (1997) and Davis (1997), depletion can be appropriately calculated if one takes depletion as estimated by current rent method I (that is, current net revenue less capital depreciation less a return to capital) and subtracts from this amount a return to the mineral resource.[11]

Production Constraints and the Hotelling Assumptions

As noted earlier, current rent methods I and II calculate total asset values based on the Hotelling valuation principle, which assumes that producers face no production constraints and that the net price rises at the rate of interest. In general, producers do face production constraints, and net prices rise at less than the rate of interest. The Hotelling principle is used as a valuation tool because of its extreme simplicity; yet, as discussed above, it has been shown both theoretically and empirically to substantially overvalue mineral reserves. Cairns and Davis (1998a, 1998b) and Davis and Moore (1997, 1998) demonstrate that asset values calculated using the Hotelling principle tend to be up to twice the market values. Thus caution is necessary in using this approach to provide asset or mineral-resource values.

Because of the potential for overvaluation using the Hotelling valuation principle, BEA uses the NPV method to adjust the stock estimates from current rent method II downward. For purposes of the present discussion, BEA's approach is termed NPV variant I. As shown above in Box 3-5, this method takes the current rent method II stock values and adjusts them downward by 12 and 31 percent using the two assumed discount rates.

The replacement cost formula is based on a model that does not require the strict assumptions of the Hotelling valuation principle and implicitly takes into account the capital constraints on oil and gas production (see Cairns and Davis, 1998a). Therefore, given the appropriate value for average costs, the model is likely to yield an accurate estimate of asset values. There has been no empirical verification of Adelman's replacement cost rule for valuing the associated capital, however, so it is not possible to judge the accuracy of the BEA method for deducting the value of associated capital to obtain the value of a mineral resource. BEA might, however, consider an alternative approach (termed here replacement cost variant II) that would subtract the replacement cost of capital from the asset value as in current rent method II, rather than the value of exploration and development expenditures.

---

[11]In mathematical terms, $depletions_t = [p_t - a_t - r_t K/q_t - D_t/q_t - rV_t/q_t] \times q_t$, where the variables are as defined in Box 3-3.

Royalty and Severance Fees

The transaction price approach has the potential to yield reasonable mineral-reserve values since it is based on observed market prices that in principle account for production constraints, market discount rates, actual reserve quality, and other factors that affect the value of mineral reserves. As noted elsewhere, however, the market value of an asset depends on the liabilities attached to the asset. In the case of minerals, production often incurs royalties, severance fees, and taxes payable to third parties as production proceeds. These and other liabilities attached to current and future production reduce the observed market value of the reserve and are deducted from the asset value by the purchaser during a reserve transaction. Thus, the observed transaction value does not represent the value of the reserves, but the value of a bundle of financial and real assets and liabilities, of which the reserves are one aspect (a point illustrated above in Box 3-2).

The treatment of these costs is not clear in BEA accounts. It appears that royalty and severance taxes are included in the unit costs used to calculate net rent in valuation methods other than the transaction method for oil and gas. This treatment is inconsistent with that under BEA's transaction price method, whereby no adjustment is made for the present value of taxes and royalties. In both cases, the pre-tax-and-royalty value of the resource will be underestimated by BEA's methods.

Revaluation

Revaluation effects are an additional element of natural-resource accounting and some other augmented accounts that are not present in the current U.S. NIPA. As discussed earlier, changes in the value of reserves are composed of additions, depletions, and revaluations (see equation 3.5 in Box 3-2).

For a simple gold-reserve case, revaluations enter the equation when reserve values adjust during the accounting period to reflect unexpected price changes. For example, suppose the average price of the existing gold-reserve stock is $10 per ounce at the start of the year, then jumps to an average of $20 per ounce on December 31. The revaluation equation becomes: revaluations ($1 billion) = closing stock value ($2.019 billion) − opening stock value ($1 billion) − additions ($30 million) + depletions ($11 million). This example shows that revaluations are calculated as a residual—the change in the value of the stock through price changes that are not taken into account in the depletion and addition calculations. Given the volatile nature of mineral prices, the revaluation component is substantial, often larger than additions or depletions. Yet the revaluation term is not directly calculated; it will include any errors in calculating additions, depletions, and opening and closing stock values.

Mineral-stock revaluations caused by unexpected changes in unit prices for reserves are calculated by BEA as a residual, and therefore are also affected by the capital depreciation schedule chosen. In the BEA data, mineral-stock revaluations are usually greater than either reserve additions or depletions, implying that most mineral wealth creation or loss comes not from additions to or depletions of the mineral-reserve base, but from large mineral price changes. Several resource economists have suggested that these revaluations are important indicators of economic welfare and should be considered equivalent to investment (gross domestic capital formation).[12] For example, a small nation could in principle sell its mineral assets to a foreign producer, and hence an upward revaluation of its assets would create wealth and higher sustainable consumption for the nation. BEA does not include revaluations in the gross domestic capital formation column of its IEESA Production Account and thereby ignores this aspect of sustainable national income.

Short-Run Volatility in Price

Where the value of a mineral asset is a function of the current extracted mineral price, as in current rent methods I and II, the NPV method, and the replacement cost method, short-run volatility in mineral commodity prices makes the value of the stock of mineral assets itself a volatile series. To the extent that price movements are temporary excursions from long-run levels, these changes in stock value will show up as revaluations. Current measures of national saving do not include revaluation effects, but future measures might do so. It should be noted that the revaluation effects in mineral assets pale in comparison with the revaluation effects from security markets.

In addition, the depletion calculations depend in part on current prices and will also be affected by price volatility. For example, consider an economy that is running down its mineral reserves at a constant rate, with no reserve additions. Depletion values will depend on current mineral prices. If nominal mineral prices increase sharply in a given year, the depletion charge will also rise sharply.

The dependence of additions and depletions on current mineral prices will affect the current value or nominal value of augmented GDP if minerals are included. Sharp changes in mineral prices could also lead to a significant change in the augmented-GDP deflator or chain-weighted

---

[12]The issue of inclusion of revaluation in income is considered in Chapter 2.

price index. The volatility of prices would not lead to volatility in the constant-price or chain-weighted indexes of real output under current concepts applied in the U.S. national accounts, but it would affect those measures of sustainable income that include elements of revaluation. These effects will necessitate considerable care in interpreting movements in GDP and its components if additions and depletions are to be added to the core GDP accounts.

BEA mitigates problems of price volatility by arbitrarily using annual prices averaged over 3 years. In addition, quantity additions and depletions are in most years nearly offsetting; thus, given BEA's approach of valuing additions at the same unit price as depletions, price fluctuations will have little impact on adjusted NDP figures. Price fluctuations do impact the stock revaluations column, but these data are not currently used in current accounting measures.

Scarcity and Long-Run Price Trends

One possible use of a series showing the change in quantity and value of a nation's stock of minerals is for assessing trends in mineral scarcity. In quantity terms, increasing scarcity might be reflected in a declining constant-dollar stock of mineral resources or of some component of mineral resources. On this front, BEA is developing a constant-1987-price series for mineral stocks, shown in Figure 3-4, that is equivalent to a physical quantity series, aggregated across different mineral types on the basis of 1987 mineral prices. This graph shows that the stock of mineral assets as a whole has been roughly constant over the 1958-1991 period. This finding might be interpreted as indicating that additions have offset depletions and that concerns about the United States running out of oil and other minerals are unfounded. Figure 3-5 shows the value of stocks and changes in current prices (from Bureau of Economic Analysis, 1994b).

The constant-price stock has limited utility as an indicator of natural-resource scarcity, however. Depletion of a physical resource indicates nothing about scarcity if that commodity is becoming worthless to society, since its disappearance will have no economic consequences. (In this respect, even chain-price indexes will not produce improved indicators.) Stock measures are particularly questionable indicators for commodities that are heavily involved in international trade, which includes all major mineral commodities. For example, many countries have seen the economic value of their domestic coal stocks decline, primarily because of the availability of low-cost coal on the world market, but this is not taken as an indicator of coal scarcity.

Relative price is usually a better index of economic scarcity, with increasing relative prices indicating that a unit of the particular asset is

Billion 1987 $

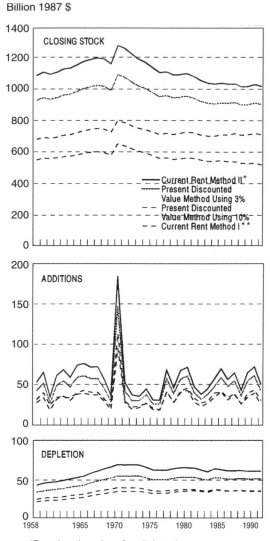

*Based on the value of capital stock.
**Based on the average return to invested capital.

FIGURE 3-4   Stocks and Changes in the Stocks of Subsoil Assets in Constant 1987 Dollars for the United States, 1958 to 1991.   Source: Bureau of Economic Analysis (1994b:Chart 2).

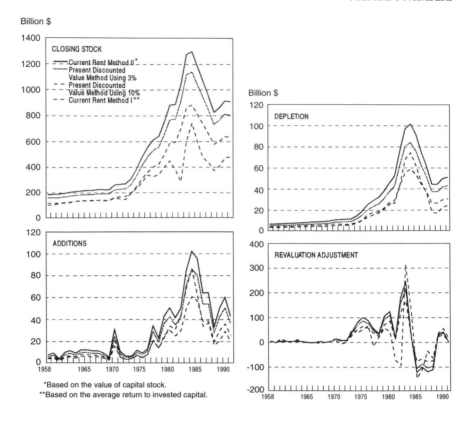

FIGURE 3-5 Stocks and Changes in the Stocks of Subsoil Assets in Current Dollars for the United States, 1958 to 1991. Source: Bureau of Economic Analysis (1994b:Chart 1).

becoming more valuable to society, and hence more scarce, relative to other assets.[13] Thus a mineral reserve's unit price is an indicator of its value to society. Increasing scarcity would be indicated by rising average reserve prices relative to other prices; for example, one might compare the relative prices of reserves and consumption goods and services or the ratio of reserve prices to the prices of other inputs, such as wage rates. These scarcity indices are not currently presented in satellite accounts. BEA does not report unit prices for reserves, and thus it is difficult to

[13]Measures of resource scarcity are reviewed in Fisher (1981:Ch. 4).

determine the implications of its findings for trends in mineral scarcity. If scarcity indicators are desired, deflated per unit prices for each type of mineral reserve should be presented.

Data Availability Issues

Although BEA's valuation methods require limited data, all may suffer from potentially significant measurement error. For example, while the replacement cost method of valuing oil and gas reserves is conceptually appropriate, it requires an estimate of the value of associated capital that cannot be measured directly and must be estimated through current exploration and development expenditures. There is no indication that this estimate, as proposed by BEA, has any empirical validity. The transaction price method is also conceptually correct, but one must make adjustments to the transactions, as listed in Box 3-2, to obtain the reserve value. The necessary data may not be available for each transaction, causing the method to lose its appeal. The current rent methods, once correctly formulated to take production constraints into account, will require average cost data that are not always observable in markets.

Other Issues

Whenever asset valuation requires discounting of future cash flows, as is the case in the valuation of mineral stocks, questions arise as to the appropriate discount rate. Finance theory offers some theoretical guidelines, but practical implementation is difficult. The popularity of the formula based on the Hotelling valuation principle derives in part from the fact that it does not require a discount rate, but this advantage comes at the cost of an implausible assumption about the increase in net mineral rents. In constructing present value estimates, it is difficult to justify the extremely low real discount rate of 3 percent per year used by BEA if the purpose of the estimates is to determine the market value of the reserves.

All NPV techniques, which include both current rent methods and the replacement cost method, omit asset value that is created by managerial flexibility (see Davis, 1996). With mineral assets, the ability to alter extraction as prices move up or down can create significant option value, especially for marginal deposits. Of the valuation techniques used by BEA, only the transaction approach includes these option values, since they will be included in the observed asset price.

BEA's results show clearly the potential margin for error among the various techniques, for they yield widely different estimates. In some cases, the net change in the value of reserves (additions minus depletions) even has a different sign under different valuation techniques. All of this suggests that correctly accounting for mineral stocks and flows in a set of satellite accounts will be just as intensive an accounting exercise as current accounting for the stocks and flows of produced capital in the NIPA.

OTHER APPROACHES AND METHODOLOGIES

Efforts in Other Countries

Mineral accounts are currently constructed by many countries. The current rent and discounted present value valuation approaches used by BEA to calculate resource stock and flow values are similar to those employed in other countries, with current rent method I being used most widely. The shortcomings of this approach were discussed earlier. Other countries assume that the current rent, after a return to capital is deducted, represents the current unit price of all reserves; they then calculate the present value by discounting the projected rent using an arbitrary discount rate. Again, as noted above, this is an unrealistic method of pricing reserve stocks or flows.

Although BEA estimates only a set of monetary accounts, most other countries compute both physical and monetary accounts for reserves. In Europe the most important minerals are oil and gas under the North Sea. Indeed, the discovery of these resources and the economic-policy problems they created led Norway to pioneer the development of resource accounting in the 1970s. Most other minerals appear to have a market value barely in excess of production costs, and hence the valuations applied to subsoil assets result in a very small value for the stocks and depletion. In Canada and Australia, however, other minerals have a significant economic value.

Coverage

The types of minerals covered in studies for other countries are similar to those covered in the IEESA. Most countries tend toward a slightly broader definition of reserves: instead of the "proven" reserves included by BEA (those that are currently known to be commercially exploitable at today's prices and technology), other countries often include "probable" reserves (defined as those having a better than 50 percent chance of being commercially exploitable in the future). Canada and Norway distinguish between "developed" or "established" and undeveloped reserves. This distinction is useful for assessing options for the future schedule of extraction. The distinction is also necessary when applying current rent method II, under which the value of associated fixed capital is deducted from the value of the reserve, and which therefore applies properly only to those reserves for which all fixed capital needed to extract the reserves is already in place.

The minerals covered by studies for other countries include oil and gas, coal, and a selection of metal ores, depending on what appears important in a given country. Hence Canada includes about 8 basic metals, while Australia values nearly 30 minerals, including precious metals and gold. In Europe, however, most minerals other than North Sea oil and gas appear to have a very small value, and efforts have not focused on them.

Valuation

The valuation methods used by other countries are generally the same as those reviewed earlier. As in the BEA work, total resource values are a small fraction of national wealth. The starting point is physical data on the stock and annual use of the minerals. As noted early in this chapter, the simplest valuation techniques are current rent methods I and II, which derive a resource rent for the current period as the difference between the extraction costs and the wellhead or surface price of the mineral. Often this margin is relatively small and can be highly volatile when the selling price of the mineral fluctuates while extraction costs undergo little change. In some cases, such as coal extraction in many parts of Europe, the mine-mouth price of coal is consistently less than extraction costs, and extraction continues only because of subsidies. A negative asset value in this case may actually be realistic.

Most countries assume that the Hotelling hypothesis is inadequate and instead use the present discounted value of the expected future income stream from extracting mineral reserves. The future schedule of extraction is often assumed to be constant, or it may actually be determined by contracts with purchasers of the mineral. In the absence of other knowledge, prices are assumed to rise with expected future inflation. The discount rate used tends to be the historical average interest rate on government bonds (typically around 6 percent), which is taken to represent the opportunity cost of funds. Normal rates of return for industry generally, or the mining industry specifically, have also been tested. Because these returns include a risk premium, they are higher than government interest rates. An interesting and quite different valuation method adopted in The Netherlands is described in the next section.

Practice in Selected Countries

Australia. The Australian Bureau of Statistics publishes values of reserves and changes in reserves for nearly 30 minerals, including oil and gas, uranium, and gold. The valuation method used is essentially BEA's current rent method I. Even in resource-rich Australia, the reported value of subsoil assets is only one-tenth the value of the fixed capital in structures and equipment. The Australian Bureau of Statistics notes that economically exploitable reserves are only a very small proportion of the total resource. It also points out that its valuation techniques can give a misleading impression both of the value of reserves and of year-to-year changes in reserves because mineral prices fluctuate considerably.

Canada. Statistics Canada has estimated the value of reserves of oil, gas, coal, and eight metals using both current rent methods I and II, although its preferred valuation technique is the latter. Current rent method I sometimes produces negative values for mineral reserves. Because Canada is concerned with regional depletion issues, it produces monetary and physical accounts for each province.

The Netherlands. Statistics Netherlands estimates the value of gas under the North Sea, the country's principal natural resource, by an unusual method. In all North Sea operations, governments (United Kingdom, Norway, The Netherlands) attempt to appropriate most of the resource rent through royalties and taxes. Instead of estimating the resource rent indirectly by the methods employed elsewhere, the Dutch estimate the resource rent directly from known government receipts. Tests by other countries have shown this method performs reasonably well for the North Sea fields, where governments take 80 percent or more of the resource rent.

Norway. The first work on resource valuation was done in Norway in the 1970s, when North Sea oil suddenly appeared as a major influence on the Norwegian economy. The Norwegians were pioneers in natural-resource accounting, beginning with oil, but later extending to other assets, such as forests. Their studies have had a considerable effect on subsequent work in other countries. The 1970s was, however, a period of massive changes in world oil prices that produced huge swings in the apparent value of this resource; as a result, many Norwegians concluded that their estimates had serious shortcomings. A number of Norwegian analysts concluded that physical data on resources were more useful. Norway recently resumed valuing natural resources to complete the balance sheets of national wealth for SNA national accounts.

Sweden. For its national accounts balance sheets, Statistics Sweden has calculated reserves and depletion of subsoil assets, in particular metal ores. The reserves covered are proven reserves, which are valued by BEA's current rent method I. Because prices of metals are volatile, the calculated resource rents occasionally turn negative, a problem reduced but not removed by adopting a moving average of prices. As a result of a fall in world copper prices, a proportion of the country's mineral stock has ceased to be economically exploitable and therefore may disappear from proven reserves.

United Kingdom. Estimates of the depletion of U.K. oil and gas in the North Sea were published in 1996 for several successively broader categories of resources—proven, probable, possible, and undiscovered but inferred from geological evidence. Several valuation techniques were tested, including current rent methods similar to those of BEA and the present value of the future income stream. Significant differences were observed in the estimates derived with the various techniques.

Other countries. Valuation studies by developing nations including Brazil, China, and Zimbabwe have produced other important findings (see Smil and Yshi, 1998; Young and Seroa da Motta, 1995; and Crowards, 1996).

Alternative Methodologies

One quite different methodology has not been employed by BEA— that of relying on financial information for individual firms. At the level of the firm, the value of mineral reserves can be imputed from data on financial balance sheets. Figure 3-6 indicates the calculations required. This method calculates a nation's mineral wealth by aggregating the values of the domestic mineral resources held by all resident mineral firms. This is a laborious process that requires assessing the balance sheets of both listed and unlisted companies. It also provides only private reserve values, since the owners of the reserve implicitly deduct the value of any taxes, royalties, and other payments on the mineral assets when attaching a value to equity capital. Finally, as with any calculation of the value of the reserve stock, it is difficult to apportion changes in total values of the mineral reserves among additions, depletions, and revaluations.

A much simpler approach entails empirically based modifications to current rent method II. Cairns and Davis (1998a, 1998b) have found that multiplying the total asset value as calculated using current rent method II by a fixed fraction can eliminate the upward bias in total reserve value and produce estimates that are closely aligned with the observed market values of mineral assets. The fraction used, which lies between zero and one, varies by commodity. Cairns and Davis' work suggests a fraction of 0.7 for gold reserves. Work by Adelman suggests a fraction of 0.5 for oil and gas reserves. For other mineral reserves, the appropriate fractions have yet to be determined, but are likely in most instances to be around

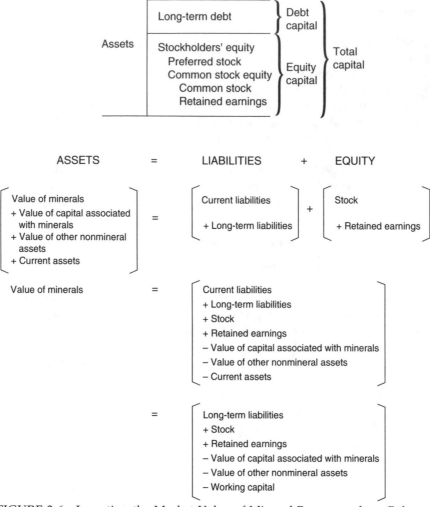

FIGURE 3-6   Imputing the Market Value of Mineral Resources from Balance Sheet Data.

0.6 according to Cairns and Davis (1998b). To estimate the value of the mineral reserves, the value of associated capital must still be deducted from the total asset value. This can be done in the same manner as in current rent method II. The mathematical formulation of this modified reserve valuation approach is shown in Box 3-9.

> **Box 3-9**
> **Modified Formulas for the Calculation of Reserve Stocks,**
> **Additions, and Depletions**
>
> total mineral reserve value$_t$ = V_t = $[\beta p_t - \beta a_t - K_t/R_t] \times R_t$
>
> additions$_t$ = $[\beta p_t - \beta a_t - Z_t/A_t] \times A_t$
>
> depletions$_t$ = $[p_t - a_t - rK_t/q_t - D_t/q_t - rV_t/q_t] \times q_t$
>
> where β is an empirically estimated adjustment coefficient with a value between zero and one, and all other variables are as defined in Boxes 3-3 and 3-6.

Additions are simply the value of new reserves, which can be calculated with the same formula used for valuing total reserves, except that exploration and development expenditures, rather than existing associated capital, are deducted. The formula for valuing additions is given in Box 3-9.

Depletion calculations have been studied by Cairns (1997) and Davis (1997), who suggest a modification to the BEA depletion calculations (see Box 3-9). Cairns and Davis take the depletion calculation of current rent method I and deduct an additional term that reflects a return to the mineral. This modification lowers the depletion calculation of current rent method I.

The discussion thus far has been aimed at estimating the value of the reserve stock and the value of depletions from and additions to that reserve stock. The discussion is guided by the notion that produced capital and natural capital are currently treated asymmetrically in national accounting and that this discrepancy should be corrected. There are yet other approaches that take a "sustainability" perspective. El Serafy (1989) has devised an alternative approach to adjusting NDP to account for mineral depletion. As currently measured, NDP is temporarily augmented during mineral extraction. El Serafy would convert the temporary revenue stream from mineral extraction into the equivalent infinite income stream, likening this latter stream to permanent income from the mineral asset. He thus advocates deducting an amount from the conventionally measured NDP during the extraction period to create an adjusted sustainable NDP.[14] It may be noted that the production of satellite ac-

---

[14]The deduction proposed by El Serafy is $R/(1 + r)^{n+1}$, where R is the current depletion, r is an appropriate discount rate, and n is the number of years of mineral reserves remaining assuming a constant extraction path. See also Hartwick and Hageman (1993) and Bartelmus (1998)

counts is intended to address just this type of concern, since those who prefer El Serafy's concept of sustainability to other accounting conventions can make their own adjustments to national output using the information contained in satellite accounts.

CONCLUSIONS AND RECOMMENDATIONS ON ACCOUNTING FOR SUBSOIL MINERAL RESOURCES

Appraisal of BEA Efforts

3.1 BEA should be commended for its initial efforts to value mineral subsoil assets in the United States.

At very limited cost, BEA has produced useful and well-documented estimates of the value of mineral reserves. These efforts reflect a serious and professional attempt to value subsoil mineral assets and assess their contribution to the U.S. economy. The methods employed by BEA are widely accepted and used by other countries that are extending their national income accounts.

3.2 The panel recommends that work on developing and improving estimates of subsoil mineral accounts resume immediately.

As a result of the 1994 congressional mandate, BEA was forced to curtail its work on subsoil assets. Its estimates of subsoil mineral assets are objective, represent state-of-the-art methodology, and will be useful for policy makers and analysts in the private sector.

3.3 Because of the preliminary nature of the BEA estimates, as well as the potential volatility introduced by the inclusion of mineral accounts, the panel recommends that BEA continue to present subsoil mineral accounts in the form of satellite accounts for the near term.

Once the accounting procedures used for the mineral accounts have been sufficiently studied and found to be comparable in quality to those used for the rest of the accounts, it would be best to consider including the mineral accounts in the core GDP accounts. It is appropriate that assessments of changes in subsoil assets be presented on an annual basis, as BEA has done in its initial efforts.

3.4 The panel does not recommend that a single approach to mineral accounting be selected at this time.

No single valuation method has been shown to be free of problems. Thus BEA should continue to employ a variety of valuation methods, modifying them as warranted by new developments in the field.

3.5 The panel has identified a number of shortcomings in current valuation approaches, and it recommends that BEA consider modifying or eliminating some of its procedures in light of these findings.

The panel has identified problems involving appropriate adjustment of asset values for associated capital and other assets and liabilities, as well as potential overestimation of the value of assets, additions, and depletions by use of the Hotelling valuation technique. BEA should consider such findings in refining its techniques. Empirically based modifications to the Hotelling valuation technique along the lines suggested above should be examined.

3.6 The derivation of accurate and parsimonious valuation is an area of intensive current research, and BEA should follow new developments in this area.

The panel has identified a number of promising research efforts that may reduce the uncertainties among various approaches to valuing mineral resources. Most of the shortcomings of BEA's approaches identified in this chapter reflect data limitations and inherent problems that arise in estimating quantities and values that are not reflected in market transactions. Given the uncertainties involved, as well as the small share of total wealth represented by subsoil assets in the United States, a major commitment to data generation for these assets does not appear to be justified at this time. BEA should therefore emphasize valuation methods that rely on readily available data.

3.7 The most important open issues for further study are (1) the value of mineral resources that are not reserves, (2) the impact of ore-reserve heterogeneity on valuation calculations, (3) the distortions resulting from the constraints imposed on mineral production by associated capital and other factors, (4) the volatility in the value of mineral assets introduced by short-run price fluctuations, and (5) the differences between the market and social values of subsoil mineral assets.

One of BEA's most important contributions has been to stimulate discussion and research on resource-valuation methodologies. BEA's actual findings regarding the value of reserves—stocks, depletions, and additions—should be considered preliminary and tentative until there is a better understanding of the magnitude of the distortions introduced by the various techniques. It is recommended that close attention be paid to these five important open issues.

Implications for Measuring Sustainable Economic Growth

3.8 The initial estimates of the subsoil mineral accounts have important implications for understanding sustainable economic growth.

In one sense, the major results of the initial estimates are negative. Perhaps the most important finding is that subsoil assets constitute a relatively small portion of the total U.S. wealth and that mineral wealth has remained roughly constant over time. According to the IEESA results, the value of mineral resources is between 3 and 7 percent of the tangible capital stock of the country. If other assets, particularly human capital, were considered, mineral value would be an even smaller fraction of the country's wealth. This is an important and interesting result that was not well established before BEA developed its subsoil mineral accounts.

3.9 Alternative measures, along with measures of sustainability from a broader set of natural-resource and environmental assets, will be necessary to obtain useful measures of the impact of natural and environmental resources on long-term economic growth.

The mineral accounts as currently constructed are of limited value in determining the threat to sustainable economic growth posed by mineral depletion. The value of subsoil mineral assets in the United States could fall because much cheaper sources of supply are available abroad. Conversely, the value could rise because serious depletion problems are driving mineral prices up. The real prices of individual mineral commodities provide a more direct and appropriate measure of recent trends in resource scarcity than is offered by the total values of specific minerals in the mineral accounts.

3.10 The panel recommends that BEA maintain a significant effort in the area of accounting for domestic mineral assets.

While subsoil assets currently account for only a small share of total wealth in the United States and do not appear to pose a threat to sustainable economic growth at present, this situation could change in the future. A good system of accounts could address the widespread concern that the United States is depleting its mineral wealth and shortchanging future generations. By properly monitoring trends in resource values, volumes, and unit prices, the national accounts could identify the state of important natural resources, not only at the national level, but also at the regional and state levels. Better measures would also allow policy makers to determine whether additions to reserves and capital formation in other areas are offsetting depletion of valuable minerals. Development of reserve prices and unit values would help in assessing trends in resource scarcity. Comprehensive mineral accounts would provide the information needed for sound public policies addressing public concerns related to mineral resources.

3.11 Efforts to develop better mineral accounting procedures domestically and with other countries would have substantial economic benefit for the United States.

Other countries and international organizations are continuing to develop accounts that include subsoil assets and other natural and environmental resources. The United States has historically played a leading role in developing sound accounting techniques, exploring different methodologies, and introducing new approaches. A significant investment in this area would help improve such accounts in the broader world economy. Unfortunately, the United States has lagged behind other countries in developing environmental and natural-resource accounts, particularly since the 1994 congressional mandate suspending those efforts.

3.12 To the extent that the United States depends heavily on imports of fuels and minerals from other countries, it would benefit from better mineral accounts abroad because the reliability and cost of imports can be forecast more accurately when data from other countries are accurate and well designed.

International development of sound natural-resource accounts would be particularly useful for those sectors in which international trade is important. Indeed, as has been learned from cataclysmic events in financial markets such as the Mexican peso crisis of 1994-1995 or the financial crises of East Asian countries in 1997-1998, the United States suffers when foreign accounting standards are poor and is a direct beneficiary of better accounting and reporting abroad. Better international mineral accounts would help the nation understand the extent of resources abroad and the likelihood of major increases in prices of oil and other minerals such as those of the 1970s. Improved accounts both at home and abroad would help government and the private sector better predict and cope with the important transitions in energy and materials use that are likely to occur in the decades ahead.

4

Accounting for Renewable and Environmental Resources

 The previous chapter reviewed issues involved in extending the national accounts to include subsoil assets. This chapter focuses on two other aspects of environmental accounting: renewable and environmental resources. BEA has proposed covering these two categories of resources in future work on integrated accounting. As discussed in Chapter 1, Phase II of that work would focus on different classes of land (e.g., agriculture, forest, and recreation land), on timber, on fisheries, and on agricultural assets such as grain stocks and livestock. Phase III would address environmental resources, including, for example, air, uncultivated biological resources, and water.

The general principles set forth in Chapter 2 indicate that increasingly severe obstacles are likely to arise as the national accounts move further from the boundaries of the market economy. The discussion in this chapter confirms the premise that BEA's Phase III raises the most difficult conceptual, methodological, and data problems. This finding presents a dilemma that must be faced in expanding the accounts: Should follow-on efforts focus on those resources that can be most easily included given existing data and methods, or should BEA focus on including those resources that would have the largest impact on our understanding of the interaction between the U.S. economy and the environment? The panel's investigation, while based on data that are highly imprecise and in some cases speculative, suggests that the development of the accounts proposed for Phase III would be likely to encompass the most significant economy-environment interactions. This observation is tempered by the realization that to date nothing approaching adequate comprehensive environmental accounting for a country of the complexity of the United States has yet been undertaken. For BEA or the federal government to prepare a full set of environmental accounts would require a substantial commitment.

This chapter provides a review of the issues involved in accounting for renewable and environmental resources. It is not intended to be a comprehensive review of work in this area. Rather, it delineates the issues that are involved in environmental accounting and presents two important specific examples that illustrate these issues. The first section reviews BEA's efforts in environmental accounting to date. Next, we analyze how stocks and flows of residuals from human activities relate to natural sources of residuals, natural resource assets, stocks, flows, and economic activity. The third section examines issues involved in accounting for renewable and environmental resources. The chapter then turns to general issues associated with the physical data requirements of environmental accounting and with valuation. We next investigate in greater detail the cases of forests and air quality to illustrate how augmented accounting might actually be done. The chapter ends with the panel's conclusions and recommendations in the area of accounting for renewable and environmental resources. Appendix B identifies potentially useful sources of data for developing supplemental accounts identified by the panel in the course of its investigation.

BEA EFFORTS TO DATE IN ACCOUNTING FOR RENEWABLE AND ENVIRONMENTAL RESOURCES

This section reviews BEA's initial design for its supplemental accounts for natural-resource and environmental assets. A more complete evaluation of BEA's efforts on forests is included later in the chapter. As discussed in Chapter 2, a critical issue involved in the development of augmented accounts is setting the boundary. How far from the boundary of the marketplace should the purview of the environmental accounts extend? Table 4-1 shows BEA's tentative decisions on how it proposed to structure its supplemental accounts (BEA's Integrated Environmental and Economic Satellite Accounts [IEESA] from Bureau of Economic Analysis, 1994a:Table 1). Phase II of BEA's development of supplemental tables focused on assets listed in rows 22-35 and 42-47 of Table 4-1, while Phase III considers rows 48-55. Because BEA has not completed Phases II and III, actual decisions on what will be included have yet to be made. Each of the following sections of this chapter considers an element of how to draw the line. While an ideal set of accounts would contain "everything,"

TABLE 4-1   IEESA Asset Account, 1987 (billions of dollars).   This table can serve as an inventory of the estimates available for the IEESA's.   In decreasing order of quality, the estimates that have been filled in are as follows: For made assets, estimates of reproducible tangible stock and inventories, from BEA's national income and product accounts or based on them, and pollution abatement stock, from BEA estimates (rows 1-21); for subsoil assets, the highs and lows of the range based on alternative valuation methods, from the companion article (rows 36-41); and best available, or rough-order-of-magnitude, estimates for some developed natural assets (selected rows 23-35 and 42-47) and some environmental assets (selected rows 48-55) prepared by BEA. The "n.a."—not available—entries represent a research agenda.

			Change			
Row	Opening Stocks (1)	Total, Net (3+4+5) (2)	Depreciaton, Depletion, Degradation (3)	Capital Formation (4)	Revaluation and Other Changes (5)	Closing Stocks (1+2) (6)
PRODUCED ASSETS						
Made assets						
1	11,565.9	667.4	-607.9	905.8	369.4	12,233.3
2 Fixed assets	10,535.2	608.2	-607.9	875.8	340.2	11,143.4
3 Residential structures	4,001.6	318.1	-109.8	230.5	197.4	4,319.7
4 Fixed nonresidential structures and equipment	6,533.6	290.1	-498.1	645.3	142.9	6,823.7
5 Natural resource related	503.7	23.1	-19.2	30.3	12.0	526.8
6 Environmental management	241.3	8.4	-7.0	10.6	4.7	249.6
7 Conservation and development	152.7	3.6	-4.4	5.3	2.7	156.4
8 Water supply facilities	88.5	4.8	-2.5	5.3	2.0	93.3
9 Pollution abatement	262.4	14.7	-12.2	19.7	7.3	277.1
10 Sanitary services	172.9	12.8	-5.6	13.7	4.8	185.8
11 Air pollution abatement and control	45.3	.6	-4.1	3.5	1.3	45.9
12 Water pollution abatement and control	44.2	1.3	-2.5	2.6	1.2	45.5
13 Other	6,029.9	267.0	-478.9	615.0	130.9	6,296.9

Item	No.						
Inventories	14	1,030.7	59.3	....	30.1	29.2	1,090.0
Government	15	184.9	6.8	....	2.9	3.8	191.7
Nonfarm	16	797.3	62.4	....	32.7	29.7	859.7
Farm (harvested crops, and livestock other than cattle and calves)	17	48.5	-9.9	....	-5.5	-4.4	38.6
Corn	18	10.2	.3	....	-1.1	1.4	10.5
Soybeans	19	5.0	-.1	....	-1.0	.9	4.9
All wheat	20	2.6	0.0	....	-.2	.2	2.6
Other	21	30.7	-10.1	....	-3.2	-6.9	20.6
Developed natural assets	22	n.a.	n.a.	n.a.	n.a.	n.a.	n.a.
Cultivated biological resources	23	n.a.	n.a.	n.a.	n.a.	n.a.	n.a.
Cultivated fixed natural growth assets	24	n.a.	n.a.	n.a.	n.a.	n.a.	n.a.
Livestock for breeding, dairy, draught, etc.	25	n.a.	n.a.	n.a.	n.a.	n.a.	n.a.
Cattle	26	12.9	2.0	....	-.3	2.3	14.9
Fish stock	27	n.a.	n.a.	n.a.	n.a.	n.a.	n.a.
Vineyards, orchards	28	2.0	.2	....	0.0	.2	2.2
Trees on timberland	29	288.8	47.0	-6.9	9.0	44.9	335.7
Work-in-progress on natural growth products	30	n.a.	n.a.	n.a.	n.a.	n.a.	n.a.
Livestock raised for slaughter	31	n.a.	n.a.	....	n.a.	n.a.	n.a.
Cattle	32	24.1	7.5	....	0.0	7.5	31.6
Fish stock	33	n.a.	n.a.	n.a.	n.a.	n.a.	n.a.
Calves	34	5.0	.9	....	-.5	1.4	5.9
Crops and other produced plants, not yet harvested	35	1.8	.3	....	.1	.2	2.1
Proved subsoil assets	36	270.0-1066.9	57.8-116.6	-16.7-61.6	16.6-64.6	58.0--119.6	299.4-950.3
Oil (including natural gas liquids)	37	58.2-325.9	-22.5-84.7	-5.1--30.6	5.8-34.2	-23.1--88.3	35.7-241.2
Gas (including natural gas liquids)	38	42.7-259.3	6.6-57.2	-5.6--20.3	4.1-14.9	8.1--51.8	49.4-202.2
Coal	39	140.7-207.7	2.2-3.4	-5.4--7.6	4.4-6.3	3.2--2.1	143.0-204.2
Metals	40	(*)-215.3	67.2-29.5	-.2--2.2	2.2-9.2	65.2-22.5	38.5-244.8
Other minerals	41	28.4-58.7	4.3-.8	-.4--.9	.1-0.0	4.6-.1	32.8-57.9
Developed land	42	n.a.	n.a.	n.a.	n.a.	n.a.	n.a.
Land underlying structures (private)	43	4,053.3	253.0	n.a.	n.a.	n.a.	4,306.3
Agricultural land (excluding vineyards, orchards)	44	441.3	42.4	n.a.	-2.8	45.2	483.7
Soil	45	n.a.	n.a.	-.5	n.a.	n.a.	n.a.
Recreational land and water (public)	46	n.a.	n.a.	-.9	.9	n.a.	n.a.
Forests and other wooded land	47	285.8	28.8	n.a.	-.6	29.4	314.6

(continues)

TABLE 4-1 Continued

	Row	Opening Stocks (1)	Change Total, Net (3+4+5) (2)	Depreciaton, Depletion, Degradation (3)	Capital Formation (4)	Revaluation and Other Changes (5)	Closing Stocks (1+2) (6)
NONPRODUCED/ENVIRONMENTAL ASSETS							
Uncultivated biological resources	48	n.a.	n.a.	n.a.	n.a.	n.a.	n.a.
Wild fish	49	n.a.	n.a.	n.a.	n.a.	n.a.	n.a.
Timber and other plants and cultivated forests	50	n.a.	n.a.	n.a.	n.a.	n.a.	n.a.
Other uncultivated biological resources	51	n.a.	n.a.	n.a.	n.a.	n.a.	n.a.
Unproved subsoil assets	52	n.a.	n.a.	n.a.	n.a.	n.a.	n.a.
Undeveloped land	53	n.a.	n.a.	-19.9	19.9	n.a.	n.a.
Water (economic effects of changes in stock)	54	....	n.a.	-38.7	38.7	n.a.	....
Air (economic effects of changes in stock)	55	....	n.a.	-27.1	27.1	n.a.	....

n.a. = Not available

* The calculated value of the entry was negative.

Note: Leaders (....) indicate an entry is not applicable.

Source: Bureau of Economic Analysis (1994a) *Survey of Current Business*, April 1994. The table has been slightly simplified for this report.

this chapter examines practical issues that arise in constructing actual accounts based on available data and tools. As will be seen, the practical is likely to fall far short of the ideal.

Pollution Abatement and Control Expenditures

One particular entry in the environmental accounts—pollution abatement and control expenditures—has been the subject of detailed investigation by BEA for many years. These items are shown for 1987 in rows 5-12 of Table 4-1. The Bureau of the Census began collecting these data and BEA reporting them in 1972 (with some breaks in the series); these efforts were suspended in 1995 because of budget cuts. Reporting of these costs does not extend the accounts, but rather reorganizes the existing accounts to provide a better indication of the interaction between the environment and the economy.

The limitations of these data are well recognized and were discussed in Chapter 2. Many of the costs included in the data overstate the cost of pollution control, while other pollution-reducing costs are omitted because they involve changes in processes. There is also controversy about the extent to which stringent pollution control regulations may have a chilling effect on innovation and technological change. Finally, little thought has been given to the appropriate treatment of purchases of emission permits, which are likely to become a more important feature of environmental regulation in the future. Despite their limitations, however, data on pollution abatement are likely to be among the most precise of the data in the environmental accounts, and they have been extremely useful for understanding trends and levels in control costs and for examining how environmental programs have affected productivity. The panel finds that the data on pollution abatement expenditures are valuable and, as noted in the final section of this chapter, recommends that funds be provided to improve the design and recommence collecting these data.

Other Sectors of the Proposed Accounts

As reported by BEA, the quality of actual entries in published supplemental accounts for Phase II and III assets ranges from relatively good to conceptually defective.[1] For Phase II assets, estimates within the category "developed land" are described as "of uneven quality" (p. 45). According to BEA, agricultural land values are "relatively good and are based on U.S. Department of Agriculture estimates of farm real estate

---

[1]All quotations in this section are from the Bureau of Economic Analysis (1994a).

values less BEA's estimates for the value of structures" (p. 45). BEA has not attempted to estimate the value of recreational land, but has entered federal maintenance and repair expenditures as an investment (see Table 4-1) and "assumed that these expenditures exactly offset the degradation/depletion of recreational land" (p. 45). BEA indicates that this assumption is made only for purposes of illustration and is "not to imply any judgment about the true value of degradation/depletion" (p. 45). A more detailed discussion of BEA estimates for timber and land in forests is presented later in this chapter.

For Phase III assets, BEA has entered "n.a." for most of the items, indicating that these estimates have not yet been developed. Entries for investment in and degradation of water, air, and undeveloped land are included, however. As in the case of developed recreational land, BEA has assumed that maintenance exactly offsets degradation, noting that this assumption provides entries that "are simply place markers" (p. 46). In the panel's view, the use of maintenance expenditures as degradation costs is highly misleading, and this procedure should not be followed in the future. Entering "n.a." would be more accurate. The panel notes, however, that these estimates do not necessarily reflect BEA's planned approaches, but were included by BEA to show the current state of data and research.

Regarding future plans, the United Nations System of Integrated Environmental and Economic Accounting (SEEA) "does not recommend that the stock of air—which is truly a global common—or water be valued; instead it recommends that valuation be limited to changes in these assets—their degradation and investments in their restoration" (p. 46). It should be emphasized that the entries for environmental assets in Table 4-1 are highly oversimplified. Some components of air quality, such as greenhouse gases and stratospheric ozone, are truly global assets and services; others, such as reductions in urban smog, are local and regional public goods. Additional dimensions that need to be incorporated are relations to external events, spatial resolution, and nonlinearities in damages. The discussion of air quality later in this chapter illustrates its multiple dimensions. Similarly, water quality and quantity, undeveloped land, and uncultivated biological resources are composites of many different assets and quality characteristics that provide multiple goods and services.

BEA's efforts have focused on the asset accounts. A preliminary table for a production account without entries is included in BEA's report on its development of the IEESA (Bureau of Economic Analysis, 1994a, 1994b). Production of market goods and services from these natural assets—e.g., timber, agricultural crops, fish—is already in the core production accounts. Greater attention is needed to identifying, measuring, and valuing the specific types of nonmarket goods and services produced by these assets.

POLLUTANT EMISSIONS AND THEIR RELATION TO STOCKS, FLOWS, AND ECONOMIC ACTIVITY

Before constructing environmental accounts, it is necessary to determine the interactions between natural resources and the environment and economic activity. It is essential to understand the key physical flows and stocks and how they affect humans and economic activities and values. A complete accounting requires detailed knowledge of the physical properties of resources and pollutants as described in fate, transport, and impact or damage models, as well as the service flows to market and nonmarket sectors.

Figure 4-1 illustrates key relationships among emissions, stocks of pollutants, natural- resource assets, and economic activities in different sectors. As the figure shows, economic activities produce a variety of uninternalized emissions and residuals that find their way into the environment. Many of the pollutants of concern are residuals that also have natural sources—sulfur, carbon dioxide, carbon monoxide, nitrogen compounds—and are emitted during volcanic eruptions, produced by forests and wetlands, or released from wildfires. Other residuals of concern— such as chlorofluorocarbons (CFCs) and many pesticides used in agriculture—are anthropogenic and have no natural sources. In terms of effects on human activities, the sources of the residuals are not important. What may be important is that human activities have increased the levels occurring in the environment, concentrated them to a degree that makes them dangerous, or relocated them to areas where people or economic activities are exposed to them at high levels.

Whether from natural sources or human activities, environmental variables can affect economic well-being in three general ways, as illustrated in Figure 4-1: (1) direct effects on consumption or income of households, industry, and government; (2) accumulation in the environment of stocks of residuals that then affect economic activities or economic assets; and (3) effects on the service flows of economic assets (capital stock, natural resources, or human resources), such as recreation, clean air to breathe, and navigable river channels free of sedimentary deposits.

Direct Effects

Environmental variables affect human and natural systems directly. Urban smog, whose concentrations change daily or even hourly, is an obvious example. Sulfate and nitrate aerosols, pollutants contributing to acid precipitation, remain in the atmosphere for a matter of days. These

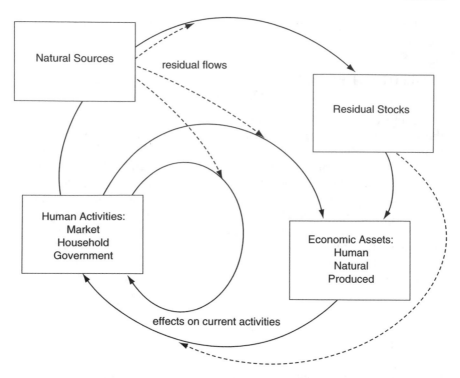

FIGURE 4-1   Human Activities, Residuals, and Economic Assets.

pollutants have short-term health effects, reduce visibility, interfere with recreational activities, affect crop growth, and present their own set of problems for accounting. In many cases, the substances emitted are precursor emissions; that is, they react chemically in the atmosphere with other substances to form the substance that is ultimately damaging to humans or ecosystems. There are also complex nonlinearities because the formation of the damaging substance depends on the level of precursor emissions, weather conditions, and the presence of other substances with which the precursor emissions react. All of these processes vary on an hourly, daily, and seasonal basis. Emissions, concentrations, and impacts of damaging substances also vary spatially, and there may be important threshold effects as well. Above all, there is the "weed syndrome"—the fact that the same substance may be beneficial or harmful depending on where it is, how much of it there is, the time and duration of exposure, and what organism is absorbing it. Virtually every substance on earth, from water to plutonium, can be an economic good or an economic weed depending on the circumstances.

One of the most important difficulties is that the physical measurements used are often inaccurate indicators of actual human exposures. Average emissions of the precursor pollutant, average concentrations over the year, or concentration data for limited sites are generally not representative of concentrations to which the population is exposed and may be a misleading basis for developing damage estimates. For example, tropospheric ozone forms mainly in warm weather. Thus total annual hydrocarbon emissions, the precursor to tropospheric ozone, are a poor indicator of potential levels of tropospheric ozone. Tropospheric ozone levels also very significantly over the distance of a few city blocks. One of the major challenges both for better environmental policy and for the construction of environmental accounts is to obtain better measures of direct human exposure to the important harmful substances among a representative sample of people.

Accumulation of Stocks

Many environmental problems result from the accumulation of residuals. These substances include most radiatively active trace gases, which remain in the atmosphere for decades or centuries, and many radioactive materials, which have half-lives of decades or centuries. Similarly, recovery from stratospheric ozone depletion is a process requiring years or decades. And agricultural chemicals often migrate very slowly through soils, contaminating drinking water only after several years or decades.

Environmental accounting therefore needs to develop and include appropriate methods to account for those persistent pollutants, such as heavy metals that accumulate in the environment and last for many years. Each year's emissions or production of residuals adds to the stock in the environment, and it is necessary to understand the processes by which these stocks decay or dissipate. In some cases (as with radioactive substances), those processes are easily understood, while in other cases (such as subsoil toxins or the carbon cycle), understanding the processes poses enormous scientific challenges. In the economic accounts, the stock-flow dynamics are similar to those of gross investment and depreciation of capital. While there is a conceptual similarity, however, there is no readily observable market price for these stock changes. Hence, valuation of a change in stock requires estimating the value of the impact of additions over the lifetime of the stock, accounting for dissipation, and appropriately discounting future effects. It should also be recognized that, with a few exceptions, the stocks are extremely heterogeneous, so that measuring a simple "environmental capital stock" is likely to be extremely difficult.

Effects on Economic Assets

Both short-lived and long-lived residuals can affect economic activity over a number of years through their effects on other economic assets, in particular produced capital goods such as buildings and equipment. For example, acid precipitation can cause deterioration of buildings. Accumulated greenhouse gases can result in coastal flooding and higher storm surges, thereby adversely affecting the value of existing coastal structures. Pollutants such as lead can cause long-lasting health consequences, impacts on intellectual functions, and premature death.

ISSUES INVOLVED IN ACCOUNTING FOR RENEWABLE AND ENVIRONMENTAL RESOURCES

The previous section addressed the major ways in which natural resources and the environment interact with economic activity. Depending on the intended uses of the data, there are different approaches to structuring environmental and natural-resource accounts. The most complete accounting structure would treat all the relationships in Figure 4-1. However, constructing such a complete set of accounts is infeasible today, and governments must choose areas for investigation strategically in accordance with their national economic and environmental goals and interests. This section delineates some possible approaches to accounting for natural and environmental resources and activities.

Production and Income Accounts

A complete set of production accounts would identify all the cross-relationships among industry, household, government, and natural sources of emissions or residuals, as well as the nonmarketed current account input services provided by nature and the productive contribution of nature to final demand. Current-year activities would include production of residuals, just as traditional economic accounts include production accounts. A complete set of accounts would incorporate flows of residuals from abroad, similar to imports of goods and services. It would also be necessary to calculate the "price"—negative or positive—indicating whether the effect was adverse or beneficial. The accounting for current-year activities would include final uses of residuals, identifying effects on final consumption, flows abroad, and contributions to capital stocks, just as traditional accounting frameworks identify final consumption of goods and services, exports, and gross capital accumulation.

Accounting for Capital Assets

It is important to measure the volumes and values of the nation's natural assets for many reasons. One purpose is simply to determine general trends. Another, illustrated in Table 4-1, is to determine the relative magnitudes of different assets. A further reason arises in the context of sustainable economic growth. As discussed in Chapter 2, one can calculate measures of sustainable income if one corrects conventional measures of national income by including the value of the change in the stocks of natural and other assets.

For all of these reasons, we would ideally like to have measures of the value and volume of the nation's natural assets; thus we must include measures not only of "made assets," such as houses and computers, but also renewable resources, such as timber or the fertility of land, and non-renewable assets, such as oil and mineral resources. It is important to know whether the economy is generating an ever-growing stock of damaging environmental residuals that will pose a large economic burden on future generations. We want to know whether the economic value of investments in tangible, human, and technological capital is more than offsetting whatever depletion of natural assets is occurring.

There is a close connection between the production accounts and the asset accounts (see Chapter 2). As noted above, measures of comprehensive income or of sustainable income include not only current consumption flows, but also the value of the change in the stocks of assets. Hence augmented accounting requires careful and accurate measurement of both assets and consumption flows. Such measurement is currently undertaken within the boundary of the marketplace, but augmented accounting would require extending that boundary for both assets and consumption in a consistent manner. The conceptual basis for asset valuation in environmental accounts parallels closely that in the conventional accounts. Depletion and degradation of natural resources is conceptually similar to depreciation of produced capital assets. Stocks of residuals can decay or dissipate, a process that is again conceptually similar to depreciation of produced assets. Natural growth of biological resources, recharge of groundwater resources, and accumulation of residuals are conceptually similar to gross capital formation or investment. Net accumulation of assets is equal to the value of the change in stocks. Many of the issues involved in constructing chain indexes of values and volumes translate directly into measurement of resource and environmental stocks.

However, some special conceptual difficulties arise in measuring stocks of natural assets. Natural-resource assets (like a physical plant or piece of equipment) are complex systems of component parts that have value because of the way they work together. Since produced capital assets are generally purchased or constructed as modules, they can be valued on the basis of their own market prices, rather than their synergistic contribution to output. To take an analogy, a baseball player's contribution to the team is a complex function not only of hitting, pitching, and fielding, but also of temperament, teamwork, and verbal abilities; from an accounting perspective, however, the economic contribution is simply wages and other compensation. For environmental assets, determining the value will become difficult when the effort extends beyond the market boundary. Consider a forest. How can the value of the stumpage in the forest be separated from the forest's contribution to erosion control, air quality, and biodiversity?

Even when markets produce evidence of the value of a bundle of assets—the composite value of soils, timber, nearness to water, and recreation—it may be difficult to separate out the values of the different components without applying complicated statistical procedures. Sometimes, the separation is misleading, as when the value of the components depends on their being together. An assembled bicycle is different from a pile of parts; similarly, forests, lakes, rivers, farmland, and coastal estuaries are valuable because of the way they are assembled.

One possible way of avoiding this difficulty is to redefine assets in terms of particular functions or characteristics, an approach similar to that taken in hedonic valuation, whereby goods are viewed as packages of characteristics. This approach would be similar to redefining an automobile as a combination of transportation mode, public-health menace, and status symbol. Under this approach, an asset is valued in terms of the sum of the values of its various characteristics. In this view, there is little point in trying to analyze the total value of holistic assets such as land or air or climate; rather, one undertakes the more modest task of looking at the different functions involved.[2] BEA's treatment of soil erosion is consistent with this approach; agricultural land is treated as the asset and the soil depth and organic-matter content as characteristics of the land. Other aspects of land quality—local climate or ambient level of pollution—can be considered in a similar manner. Identification of the economic effects of erosion on the value of land makes the resource link explicit.

Thus, a potentially useful alternative to considering the holistic value of assets is to consider how changes in air quality affect the value of agricultural land, forests, residential property, and human capital. Thus, fundamental nonhuman assets might include forests, lakes, rivers, estuaries, coastal regions, wetlands, farmland, and residential property. This

---

[2]Watershed valuation is an example of a holistic approach (see Anderson and Rockel [1991] and Green et al. [1994] as examples).

approach has two further attractive features: it allows better integration with existing accounts, since some of these assets (such as residential property and forests) have an extensive existing database; and it allows incremental development of a set of valuations, building upon those in the market sector.

Practical Choices in Expanding the Accounting Framework

A complete accounting system including interactions in the production and asset accounts would be a significant undertaking. Deciding on the scale of augmented accounting and the next steps to be taken will require considerable strategic thought. One question is whether the accounts will be used for scorekeeping or for management (see the discussion in Chapter 2).

Scorekeeping, which involves developing a better measure of the performance of the economy over time, is one perspective. It addresses the questions of trends in the values of environmental assets and whether current consumption is sustainable. If scorekeeping of this type is the purpose of supplemental environmental accounts, it will simplify the enterprise because there will be no need to consider intermediate interactions between production sectors. Tracing where pollutants were produced and how they affect intermediate product is unnecessary as long as one can measure final consumption and changes in assets. For example, a dying forest is a deteriorating asset; whether the deterioration is caused by acid precipitation, tropospheric ozone, or pest infestation is secondary from a scorekeeping perspective. What is important is to measure the deterioration accurately. Similarly, the overall health and skills of human populations is a central issue in measuring whether the economy as currently structured is leading to an increase or decrease in the stock of human capital. Why the change is occurring—whether because of changes in health care or education expenditures or reductions in blood lead—is secondary to the measurement issue. Overall scorekeeping would note the substantial improvements in the health status of Americans over this century rather than decreases in particular ailments.

The second broad perspective on the function of environmental accounts is that of environmental management. This perspective focuses on the sources, transportation, and ultimate disposal of residual pollutants, particularly their contributions to outcomes of economic and ecological consequence. Knowing to what extent particular emissions of residuals come from utilities, automobiles, or volcanic eruptions is critical to developing strategies for control. If human sources are dwarfed by natural sources, for example, efforts to control human sources may be futile. Similarly, knowing that life expectancies have increased dramatically is not very useful to understanding whether there are benefits to tightening controls on small particles or ozone. Improvements in health care, occupational safety, and traffic safety may result in increasing life spans and health status more than pollutants are shortening life span—but reducing pollution further could extend lives further. Thus, if the supplemental accounts are meant to support environmental management decisions, knowing the sources of pollutants and the specific causes of changes in asset quality are essential.

Analogy with Economic Accounts

The discussion in this section has emphasized the complexity involved in constructing environmental accounts. It is useful to compare environmental with conventional economic accounting. A little reflection suggests that economic activity has a similar, almost fractal complexity when one looks under the surface. It would be just as difficult to measure the physical flows in economic life as in environmental life, and indeed many of the same processes come into play. Consider the problems involved in accounting for a simple loaf of bread. Doing so would require measuring and valuing a wide variety of flows of water, fertilizer, pesticides, labor, climate, and capital inputs that go into producing the wheat; the fuels, transport vehicles, emissions, weather-related delays, induced congestion, or floods involved in transportation; the molds, spores, and miscellaneous rodents and their droppings that invade the storage silos; the complex combination of human skills, equipment, and structures that go into milling the wheat; the entrepreneurship of the baker and the software in the computer-operated baking and slicing machinery; the complex chemistry and regulatory environment involved in the wrapping materials; and the evolving ecology of the distribution network. Behind each of these elements, in addition, is the complex general equilibrium of the marketplace, which determines the selection of production processes by prices, taxes, and locations, along with the further complexity of needing to unravel the input-output structure of the inputs into each of the steps just described.

It appears unlikely that anyone would try, and safe to conclude that no one could succeed in, describing the physical flows involved in this little loaf of bread. Fortunately, however, economic accounting does not attempt such a Herculean task. Rather, the national accounts measure all these activities by the common measuring rod of dollars. Although the dollar flows are routinely broken down into different stages—wheat, transportation, milling, baking, and distribution—one could never hope to describe the flows physically and then attach dollar values to each physical stage. Yet this is just what would be required for a full and detailed set of environmental accounts. The above comparison may give some sense of why accounting for environmental flows outside the marketplace is such a daunting task.

PHYSICAL DATA REQUIREMENTS: GENERAL ISSUES

Some of the analytical questions involved in environmental accounting have been analyzed in the previous section. To construct actual accounts requires both obtaining accurate physical data (discussed in this section) and valuing the flows (discussed in the next section).

Accurate data on physical flows and stocks are a prerequisite for developing any accounting system and are the focus of national accounting systems under development in several European nations. In some areas, ample physical data are available as a by-product of regulatory monitoring and resource management systems. Appendix B lists a number of databases identified by the panel that may be of use in further work on supplemental accounts.

Three concerns are fundamental to understanding data and measurement requirements for the development of environmental accounts: (1) the dose-response relationship, (2) measurement of actual doses experienced, and (3) the fate and transport of residuals in the environment. The first, the dose-response relationship, is the physical relationship between the concentration of or exposure to an environmental change and the response of the subject experiencing the dose. The dose-response relationship is applied to many different situations, for example, the response of trees and crops to chemicals such as carbon dioxide, tropospheric ozone, or acid deposition and the response of humans to pollutants such as lead, particulate matter, or radiation.

Dose-response relationships are often difficult to determine because they may be affected by complex interactions and intervening factors. For example, there are extensive medical data on causes of death and, less universally, illness. To determine impacts of environmental changes on human or natural ecosystems requires separating out the different causes of premature death or illness. In some areas, such as the impact of tobacco or lead, the relationships are relatively well established; in other areas, such as the impact of particulate matter or ozone, much uncertainty persists. For many of these relationships, average exposure over the year is rarely the relevant measure. Damage may be related to extreme levels or to periods in which the subject is particularly sensitive to the agent; acute effects may differ from chronic effects related to long-term, low-level exposure.

Resolving these uncertainties about dose-response relationships is important for policy decisions, such as the level at which to set primary air-pollution standards. Resolution of these uncertainties would also allow construction of environmental accounts. The panel's review of work in this area indicates that the preparation of estimates of the economic impacts of air pollution is feasible today, but there are enormous uncertainties at virtually every stage of the effort. While BEA or those preparing environmental accounts would not necessarily be involved in preparing dose-response estimates, the accountants will need to work closely with public-health, agricultural, forestry, and ecological experts to use the best information available.

In addition to understanding the dose-response relationship, national accounting requires regular, statistically valid monitoring of the relevant populations and the doses they are receiving. A basic limitation of much of the data currently collected is that ambient concentration levels in areas where individuals, crops, forests, or other relevant entities actually reside are poorly measured. Most measurements occur at sites of convenience rather than sites of relevance. Air pollution monitors are often placed with other monitoring devices where airplanes congregate rather than where people live.

A full account of economic-environmental interactions also requires tracking the fate and transport relationship, or the connection between the emission of a particular pollutant or pollutant precursor at one time and geographic point and the level, time, and location of the pollutant at the point where it affects an economic asset or activity. These relationships are generally highly complex and variable. For air pollutants, wind direction and speed, temperature, cloudiness, and precipitation all affect how a pollutant is dispersed or concentrates. Precursor pollutants sometimes do not create damage themselves, but react chemically in the atmosphere to create other agents that are damaging. Acid precipitation and tropospheric ozone are examples. The formation of these pollutants depends on the presence of other agents that may limit, speed, or slow the process. Monitoring of emissions, concentrations, exposures, and consequences would provide the physical foundation for a complete set of environmental accounts, and is also a critical part of environmental management.

The goals of environmental accounting will dictate the assignment of priorities for improved data. Extensive data on the fate and transport of emissions and concentrations of pollutants are a lower priority if the goal is scorekeeping; even dose-response relationships may be secondary to more direct measurement of consumption flows or changes in important capital and environmental assets and human health status. If one is interested primarily in measuring the sustainability of economic activity, understanding the health status of human and natural systems is more important than understanding why conditions have changed. On the other hand, understanding these technical relationships is essential if environmental accounts are to serve as a data set to support environmental management, in which the goals are to understand the severity and causes of environmental problems, along with remedies needed to mitigate those problems.

VALUATION: GENERAL ISSUES

Once appropriate physical data have been developed, the next step in developing integrated accounts is to value changes in the physical measures. Physical data alone are often interesting and useful for policy making, and improvements in physical environmental data could enhance policy-making efforts. Indeed, most countries have not gone beyond developing physical measures and indicators because of the difficulties involved in valuing nonmarket goods. Without valuation, however, physical data alone have serious limitations for both scorekeeping and environmental management. Aggregate physical measures, such as areas of agricultural land, forest, or wetlands or tons of sulfur, toxic wastes, or particulate emissions, provide incomplete evidence on the effects of these chemicals on economic well-being or economic sustainability over time. For example, losing 1000 acres of prime Florida Everglades would probably impose a greater economic and ecological loss than losing an equivalent area of frozen wetlands in northern Alaska. Thus an accounting entry of "total wetland acres" lost would not be a useful measure. Furthermore, a simple measure of wetland area would fail to capture improvements in quality that might occur as a result, for example, of current efforts to restore the Everglades as a fully functioning ecosystem.

For many issues, it is necessary to weight the physical measures by their importance. There are approaches to weighting physical quantities other than valuing all impacts in dollar terms; for example, different environmental residuals can be weighted by how they affect human mortality. However, such weights would be incomplete because they would exclude impacts on morbidity or on the health of ecosystems. In economic accounting, the "importance weights" are the economic values, usually market prices. The advantage of using economic valuation is that comparisons can be made across very different environmental effects and with goods that are part of the market economy. While relying on economic values has many desirable features, there are a number of difficulties involved in usefully applying nonmarket valuation studies and techniques to environmental accounting, as discussed below (see also Chapter 2).

Valuation Techniques

Markets provide the conventional valuation for market goods and services. A variety of methods for valuing nonmarket goods and services has been developed. Table 4-2 indicates the potential and actual uses of various valuation methods for many environmental problems, including the dose-response method discussed above. These methods have been developed over a number of years and have been applied to many specific problems.[3]

The *dose-response method*, as a valuation method in and of itself, is directed toward converting exposure to a specified dose of a substance, from which is calculated a physical response for which a direct market price can be observed. For example, exposure to ozone or particulate matter results in wheat-yield loss or lost work-days due to respiratory illness; using the market price of wheat or of labor, an estimate of economic value can be made. The valuation techniques in this approach are consistent with prices used in the economic accounts. Incomparability or additional uncertainties are introduced only through imputation of output by use of the dose-response relationship, which converts the environmental effects into market-good terms.

Travel-cost and hedonic methods also use behavior and observed market transactions as a basis for estimating values, but the activities involve time use and expenditures on goods and services related to use of the environmental or natural-resource good, rather than on the resource itself. For example, a recreational site might be valued using the travel-cost method by estimating the time and out-of-pocket costs involved in reaching the site.

Hedonic methods use statistical techniques to explain variations in market prices based on the bundle of characteristics of a good. This approach is currently used in the national accounts. Computers, for example, are considered bundles of attributes such as speed, memory, and random access memory (RAM), and the value of the computer is a weighted sum of the values of its attributes.

For resource and environment valuation purposes, hedonic methods are used to explain variations in land values that reflect natural-resource or environmental characteristics. Such estimates are based on observed price differences of land with different amenities or disamenities such as noise, pollution, and crime. Hedonic wage studies—looking at the wage premiums of high-risk jobs—are currently the standard approach to esti-

---

[3]See Smith (1993) and Braden and Kolstad (1991) for reviews of the theory and application of these methods.

mating the value of workplace hazards; the results are often used as estimates of the value of life-threatening effects due to such causes as air pollution or traffic accidents.

Contingent value (CV) methods are survey techniques that ask people directly what they would pay for goods and services. Applications in the area of environment and natural resources include, for example, asking individuals what they would be willing to pay to reduce smog, to increase visibility in places such as the front range of Colorado, and to clean up an oil spill in a coastal area. CV methods differ from the other methods discussed above in that there are no budget constraints or behavioral observations involved; the results reflect respondents' estimates of the value of a hypothetical change, rather than a dollar or time cost actually incurred. While widely used for environmental valuation, CV is highly controversial because it often fails elementary tests of consistency and scaling and is subject to a wide variety of potential response errors if not carefully constructed.

The overriding problem with all these methods is that they require voluminous data and statistical analysis and can hardly be used routinely for a large number of products in constructing environmental accounts. Where existing CV studies are used for environmental or natural-resource valuation, they often employ valuation approaches that are inappropriate for national accounts. For example, many estimates used in environmental management rely on average value (including consumer surplus), rather than the prices or marginal values that are the convention in national income accounting.[4] In a competitive economy, market prices measure both the incremental value to the economy of consuming another unit of the good and the incremental cost to the economy of producing that unit. Therefore, prices are a useful benchmark for valuation.

In one sense, the market value underestimates the total value of goods and services to consumers. Because consumers pay the price of the last or marginal unit for all units consumed, they enjoy a surplus of total satisfaction over total cost. The term used for the extra utility consumers receive over what they pay for a commodity is *consumer surplus* (see also Chapter 2). Consumer surplus introduces a complication in comparing market prices with nonmarket values. For goods without markets, value is often measured by total willingness to pay for the good. Such values are not directly comparable to market prices because the values include

---

[4]Marginal costs and marginal values are central concepts in determining economic efficiency. For example, knowing the marginal value of reductions in atmospheric lead is more useful to the policy maker than knowing the average value of all reductions. Marginal cost and marginal value are defined in Appendix D.

TABLE 4-2 Methods for Environmental Valuation

Pollution	Type of Effect	Impact	Techniques for estimation impacts				
			Hedonic Property	Hedonic Wages	Travel Cost	Contingent Valuation	Dose Response
Air pollution							
Conventional pollutants: (total suspended particulate [TSP], sulfur dioxide [SO$_2$], nitrous oxides [NO$_x$])	Respiratory illness	WLD RAD Medical suffering	L	L	X	√	√
	Respiratory illness	Death	L and √	√	X	X	√
	Aesthetics	Visual, sensory	√	L	X	√	X
	Recreation	Visits, especially to forests	L	X	√	√	X
	Materials	Maintenance/repair	X	X	Poss	Poss	√
	Vegetation	Crop losses	L	X	X	X	√
Water pollution							
Conventional pollutants (e.g., biochemical oxygen demand [BOD])	Recreation (e.g., fishing, boating)	Visit behavior	L	X	√	√	X
	Commercial fisheries	Stock losses	X	X	X	X	√
	Aesthetics	Turbidity, odor, unsightliness	√	X	L	√	X

Category	Effect	Impact					
Trace concentrations	Ecosystem	Habitat and species loss	X	X	X	√	√
	Drinking water	Illness, mortality	X	X	X	Poss	√
	Fisheries	Stock losses	X	X	X	X	√
Toxic substances Air (benzene, polychlorinated Biphenyls [PCBs], pesticides)	Illness, mortality	WLD / RAD / Medical expenses / Pain and suffering	√	√	√	√	√
Chemicals hazardous to land	Aesthetics / Ecosystem	Unsightliness / Anxiety, ecosystem losses	X	X	X	√	√
Radiation	Illness, mortality	WLD / RAD / Lives lost	Poss	√	X	L	√
Marine pollution Oil, radioactive substances, sewage	Aesthetics / Swimming	Unsightliness / Visit behavior / Illness / Fish/livestock losses	√	X	√	√	√
Noise	Nuisance	Annoyance	√	X	X	√	L

√ = Used technique; Poss = Not developed, but possible; X = Inapplicable technique; WLD = Work loss days; L = Very limited applications; RAD = Resource activity days.

Source: Adapted from Organization for Economic Cooperation and Development (1989), as appearing in Costanza (1997).

the consumer surplus. In other words, when nonmarket goods are valued according to total willingness to pay, the value of those goods is overstated relative to the market value of marketed goods. For example, travel costs can provide the average value of a recreational service, but the marginal value of the resource for an open-access beach or forest with no fee may be zero. This discussion illustrates the importance of ensuring comparability in estimating values in the construction of nonmarket economic accounts.

Classes of Economic Goods

The valuation of environmental goods and services raises an issue that is largely overlooked in conventional accounting—the distinction between private and public goods. These deceptively common terms are used in a specialized sense here (see Samuelson, 1954, 1955). *Private goods* are ones that can be divided up and provided separately to different individuals, with no external benefits or costs to others. An example is bread. Ten loaves of bread can be divided up in many ways among individuals, and what one person eats cannot be eaten by others. *Public goods*, by contrast, are ones whose benefits are indivisibly spread among the entire community, whether or not individuals desire to purchase them. An example is smallpox eradication. It matters not at all whether one is old or young, rich or poor, American scientist or African farmer—one will benefit from the eradication whether one wants to or not. The example of smallpox eradication is a dramatic case of a public good. The economy is replete with activities, such as pollution abatement, new scientific knowledge, national defense, and zoning, that have public-good characteristics.[5]

---

[5]This discussion greatly simplifies the discussion of public goods. There are further distinctions among public goods that are central to many issues involved in environmental accounting, particularly as regards valuation methods. One such distinction is whether consumption is excludable; in the case of global warming, for example, no coastal nation can exclude itself from the rising seas. Another distinction is between pure and congestible public goods. Congestible public goods are those whose consumption is neither completely rival nor nonrival; one person using a beach does not preclude others from doing so, but most people find crowded beaches less enjoyable than deserted ones (see Cornes and Sandler, 1986). Crowding of this sort means that even with open access, the marginal value of use of these sites is greater than zero. A final distinction is between those goods whose use affects market activities or market values and those that are completely independent of the market. Public goods without traces in markets are frequently referred to as "nonuse values." Nonuse values include values people derive from knowing that a species exists, natural wonders remain, or natural systems survive intact beyond any specific use to which they might be put (see Randall and Stoll, 1983). When Congress created Yellowstone National Park in 1872, for example, no member of Congress had ever been there, and its value as a natural wonderland was largely a "nonuse value" imagined on the basis of photographs of William Henry Jackson and drawings of Thomas Moran.

The distinction between public and private goods is central for many nonmarket and environmental commodities. In a perfectly competitive market, the price of a marketed private good is the marginal value of consumption to the consumer. Similarly, while observed prices do not exist for nonmarket private goods, the marginal value of the consumption of such goods is conceptually equivalent to a market price. The national accounts value food produced and consumed on farms, even though it is not marketed, the same way food sold in the marketplace is valued.

Valuation of public goods is an especially difficult problem because their value to all consumers must be reckoned with. For example, improvements in air quality affect everyone. Conceptually, therefore, one should value public goods by adding up the marginal values of changes to the entire affected population. Doing so poses severe measurement difficulties for two reasons. First, the "personal prices" or marginal values of the public good are sure to vary across people—some may be significantly affected and therefore place a high value on air quality, while others may be relatively indifferent. Second, determining the values of public goods is extremely difficult because people make few decisions that reveal their preferences in this regard. People cannot choose how much defense or smallpox eradication they would like to consume; these decisions are made collectively. Since people cannot choose different levels of a public good, there are no behavioral traces of their preferences or personal prices.

For the above reasons, constructing environmental accounts will necessarily be different for private and public goods. For private goods, particularly near-market goods that have close relatives in the market economy, valuation appears feasible and has a level of reliability that approaches that of the current national income accounts. Most public goods, by contrast, present greater measurement and conceptual problems. Table 4-3 shows examples of each type of goods that have these different characteristics.

Strategies for Valuation

Near-market natural-resource and environmental goods (which are largely private goods) offer the most promise for valuation and inclusion in the accounts. Often there are markets for comparable goods that provide direct evidence of the value of the nonmarketed goods or services. This approach is consistent with the use of market prices used elsewhere in the accounts and has precedent in the valuation of owner-occupied housing services. Thus, the methods for including these near-market goods have already been established. A potential source of error in using this approach is that the quality may differ for goods or services pro-

TABLE 4-3   Classes of Goods and Services

Type of goods	Private (examples)	Public (examples)	
		Related to Markets	Independent of Markets
Market	Bread Cars Restaurant meals Housing rentals	Knowledge and innovations that are patented and copyrighted Pollutants with tradeable permits	None
Nonmarket	Household prepared meals Leisure time Television viewing Groundwater for drinking Rental values of owner-used assets	Air and water quality Climate Mosquito control	Passive or nonuse value (e.g., knowledge of the existence of species, unique national treasures such as Yellowstone National Park)

duced or provided in the household and those produced in the market. It would be appropriate to undertake a modest research program to investigate the adjustments necessary to make market and near-market activities comparable.

Two basic types of near-market goods are of interest. The first is the service flow from a natural resource. Here, as in the case of timber from forests or crops from farmland, the service flow is already in the core accounts, and the returns to these assets appear as profits and/or returns to other assets, but the accounting is incomplete because it omits the nonmarket activities. The second case is a good not currently in the accounts, such as recreation services enjoyed by households; in this case, the value that is attributable to the service is equal to the value of household labor and capital services, plus a service flow from a natural resource.

Public goods that affect markets offer opportunities for using observations of actual market transactions to generate valuation estimates. An example would be concessionaire activity within a national park. The hedonic property and wage techniques can be explored as a basis for developing values or imputing how changes in these public goods affect markets. There are some potentially sound ways to make the links between these public goods and the market explicit in the accounts, but there is not yet a consensus on how to include them, and each provides a challenge for data development and estimation of values.

Other classes of public goods, particularly those that are national or global in nature and do not leave behavioral traces of individual preferences, are currently problematic for the national accounts. Most of these public goods, such as those involving nonuse values of natural-resource and environmental assets, can be valued only with CV methods. Some reviews have conveyed cautious approval for use of these methods in limited circumstances. For example, a panel convened by the National Oceanic and Atmospheric Administration to review CV methods for use in federal compensation decisions identified "a number of stringent guidelines for the conduct of CV studies" that, when followed, allow "CV studies [to] convey useful information" (see Arrow et al., 1993:4610). However, the accuracy of the values developed with these methods remains controversial among those in the economics profession (see Portney, 1994; Hanemann, 1994; Mitchell and Carson, 1989; and Diamond and Hausman, 1994).

As discussed above, the hypothetical nature of the valuation makes these methods quite different from other methods that are based on actual market transactions. For these reasons, while CV is sometimes useful for other purposes, the panel has determined that it is currently of limited value for environmental accounting. This means that, for many important environmental assets, environmental accounts will omit a portion of the value of the assets. That is, it appears to be feasible to work toward accounting for goods such as recreation activities associated with the Florida Everglades, Yellowstone National Park, and similar sites. However, it is beyond the ability of current techniques to provide reliable measures of the value of the public-goods services provided by these assets, even though we may suspect that these services are precious to the nation.

In the remaining sections we explore the issues raised in the preceding sections in far more detail for the cases of forests and air quality.

FORESTS: A RENEWABLE NATURAL RESOURCE

Forests are a prime example of renewable natural-resource assets. They present many of the same national economic accounting issues as other renewable natural-resource assets, such as agricultural land, fisheries, and coastal and freshwater resources. Many of the products derived from natural-resource assets are included in the production accounts of the existing core NIPA. But these assets are not generally included in national asset accounts, and the production accounts themselves exclude many nonmarket goods and services derived from these natural-resource assets. Forests are a useful example because much effort has been devoted internationally to forest accounting.

While the NIPA as currently structured are not intended to include the full range of forest values, regular reports of economic activity as measured by the NIPA are widely noted and interpreted as measuring important aspects of economic well-being. It is logical to try to capture in these accounts more of the important relationship between forests and humans. Forests support human material and spiritual welfare in countless ways. They harbor many important species of plants and animals. They form an aesthetically pleasing backdrop for recreation and for everyday life. They filter and regulate the flow of much of the U.S. water supply. They have been a reservoir for land available for conversion to agriculture and other developed activities. Wood is one of the world's most important industrial raw materials and a ubiquitous source of energy. And worldwide, literally millions of indigenous people call forests home.

This section examines, in five parts, methodological and practical issues that arise with regard to including forests in national economic accounts. It begins with a discussion of the nature of the economics of forest values, providing a general framework for assessing those values. The second subsection translates this general discussion into a more precise statement of how forest values might be incorporated in the U.S. economic accounts. Given this context, the third subsection comments on BEA's work to date and provides a brief discussion of the extensive international literature on forest accounting. This is followed by discussion of a recommended approach for measuring the net accumulation of timber. The section ends with the panel's conclusions on forest resources.

The Nature of Forest Values

Forests produce economic value through three principal classes of economic goods: private goods traded in markets, private goods not traded in markets, and public goods. These goods can affect both the national asset accounts and the NIPA.[6] These three classes of forest goods and services are discussed in decreasing order of availability of data and of accepted analysis required to include them in the national economic accounts.

---

[6]The following discussion focuses primarily on issues pertinent to the United States. A significant issue in natural-resource accounting for many developing countries is deforestation. For example, a major concern in the national accounts of developing countries such as Indonesia is that harvesting of forests is contributing to rapid growth in current consumption at the expense of the stock of forest assets. In the late 1800s, the deforestation rate in the United States equaled or exceeded that found in many tropical countries today, but deforestation is no longer significant on a national scale, and the general trend since the 1950s has been a net growth in the forest stock of the United States.

Private, market-related activities. Some forest-based market-related activities are already included in the national income accounts; examples are all forest products used in manufacturing (logging, lumber production, the manufacture of paper, wooden furniture, and musical instruments). Some fuel wood production would fall into this category; the part that flows through the market economy would enter the accounts, while the part that is produced for own consumption would not.

The major issue in the current treatment of private, marketed forest-based goods and services is the failure to account for changes in the value of the standing timber. Most of the conceptual problems involved in doing so have been fully considered and developed, as discussed below. Accounting for changes in the timber inventory would address one of the major shortcomings of the existing forest accounts.

Private goods not traded in markets. Forests produce many private goods and services that—for reasons of custom, law, or economics—society has elected not to allocate through markets.[7] For example, the water flowing from forested watersheds has considerable economic value. Indeed, the rationale for forest conservation in the late nineteenth century related primarily to protection of forested upland watersheds. Protection of navigation was the explicit constitutional basis for creation of the eastern national forests, and congressional agricultural interests concerned about irrigation provided the principal support for withdrawing the national forests from the western public-domain lands. A study by Bowes et al. (1984) of the Front Range of the Rockies around Denver and informal estimates for the Quabbin Watershed servicing Boston demonstrate that in some locations, the value of the water produced from a forest may far exceed the value of the timber production. Changes in forest attributes can affect stream flow and therefore the value of water "produced." Interestingly, Bowes et al. (1984) demonstrate that when water is valuable, it is optimal to keep timber stocks low to reduce evapotranspiration and therefore increase runoff.

Public goods. Public goods are ones for which consumption by one individual does not reduce the amount available for others to consume. Forests produce many public goods, including aesthetically pleasing landscapes, a carbon sink, and a store of biological diversity. Given data on changes in forest inventories, it may be possible to value some of these services (e.g., the value of carbon sequestration), although the uncertain-

---

[7] Because of the decision not to use markets in allocating such resources, but typically to provide them through collective decisions, common usage sometimes refers to such goods and services as "public goods." This report follows the conventional definitions of public and private goods discussed in the previous section.

ties of such valuation should not be underestimated. In other cases, the valuation problems go far beyond the results of current research.

The interactions among these three sources of forest value—private marketed goods, private nonmarketed goods, and public goods—can be complex. For example, cutting trees leads to increases in manufacturing activity. This in turn might cause an increase in water yields and thereby reduce the costs of industrial and household production. It might also cause a shift of species diversity away from late-seral-stage organisms, such as spotted owls, and toward early-seral-stage ones, such as elk. It would lead to an immediate release of carbon associated with logging and forest products manufacturing, but might result in a long-term increase in carbon sequestration with forest growth if the wood products were sequestered in long-lived furniture or houses. Given the site-specific nature of such production relationships and the lack of current scientific understanding of many of the underlying ecological processes, there is currently an insufficient scientific basis for specifying a full set of such linkages in supplemental accounts.

Incorporation of Forest Values in the National Economic Accounts[8]

To be most useful, the economic accounts would identify the separable contributions of forests to the national economy. It is convenient to discuss the problems involved in incorporating forest values in the U.S. national economic accounts first for the production accounts and then for the asset accounts.

Adjustments to Production Accounts

A full treatment of forests in the production accounts would involve the following adjustments to national income and product.

Timber income. Sales of timber are already included, although some are recorded as part of personal income, some as part of manufacturing income, and some as part of government receipts. The principal difficulty is ascribing these income streams to the forest sector; in this respect, the issues are very similar to those encountered in the treatment of mineral incomes discussed in Chapter 3. Ordinary production costs associated with forest production activities are similarly covered by the current NIPA, but may not be easily associated with the forests themselves, rather than forest-products manufacturing. Problems remain with the alloca-

---

[8]The discussion in this section draws heavily on the recent comprehensive treatment of the subject by Vincent and Hartwick (1997).

tion of joint costs. For example, forest roads are a costly input to the production of many forest products, including timber, minor forest products, and recreation. Yet standard accounting practices, especially for the national forests, attribute the full cost of these roads to the timber program. As currently constructed, the NIPA include the costs of road construction, but exclude the benefits produced by the road.

Near-market forest products. To the extent that near-market forest products, such as fuel wood, berries, mushrooms, and Christmas trees, are produced by households but not purchased through markets, they would be included in the forest accounts.

Contributions to household production (e.g., recreation). The accounts would include the value of household production of activities such as hiking, hunting, and fishing. However, if there is uncongested, open access to the forest-based inputs needed for household production, the contribution of these inputs to household value on the margin is zero. Current practice often uses average rather than marginal values, so care must be taken, particularly for open-access forests, to ensure consistent valuation in order to prevent overvaluation of nonmarket activities.

Environmental services used by other industries (e.g., watershed protection, domestic/industrial water supply). Some of the impacts of forests are already included in the NIPA. For example, if forests moderate water flows and reduce the cost of agricultural production, this benefit is fully incorporated in the NIPA. Ascribing the benefit to the forest sector, while a difficult task, would be required for a full accounting.

Public goods (e.g., carbon sequestration, biodiversity, species preservation). At present, the only public goods that have been the subject of widespread attempts at valuation are those associated with carbon sequestration (Brown, 1996). While quantitative data on carbon sequestration are available, valuation is still highly uncertain. Moreover, because valuation of carbon sequestration is based on global benefits, the issue of how such benefits would be incorporated in a single nation's accounts is unresolved.

There are few comprehensive studies of the total value of forest products. Recent work on goods and services produced on public lands managed by the U.S. Forest Service indicates that more forestland value is due to recreational and wildlife services than to timber, mineral, and range goods (U.S. Department of Agriculture Forest Service, 1995). For example, of the estimated total $9 billion value of forest goods and services in 1993 (valued at market prices), recreational and wildlife services accounted for 80 percent, whereas the production of minerals and timber and grazing range services accounted for just 20 percent.

While the above estimates illustrate the importance of nonmarket production, they should be interpreted with caution. First, they include only land managed by the U.S. Forest Service, which is not representative of all forestland. By contrast, on private lands that are intensively managed for timber production, much of the value is due to timber harvesting. Second these estimates do not include all nonmarket values; for example, they omit the potential value of carbon sequestration. A recent estimate is that U.S. forests sequestered 211 million metric tons of carbon in 1992 (Birdsey and Heath, 1995). At $10 per ton, a value consistent with the Intergovernmental Panel on Climate Change (IPCC) estimates of the marginal value of emission reductions (see Bruce et al., 1996), the annual value of carbon sequestration in all U.S. forests would be $2.1 billion; the numbers could be an order of magnitude larger if the U.S. adopted stringent emission controls under the Kyoto Protocol of 1997. Third, the Forest Service presents different types of estimates for the value of forest services, market-clearing prices being only one of these.[9]

Forests Asset Accounting

A key conceptual problem with the present NIPA is the lack of any accounting for changes in asset values of U.S. forests. Accomplishing this task was part of the Phase II work outlined by BEA (see Chapter 2). We address this issue in some detail for two reasons. First, from a conceptual standpoint, natural-resource assets should be treated consistently with produced capital assets, adding net accumulation or subtracting net decumulation from gross domestic product (GDP) to arrive at a measure of net national product (NNP) more closely associated with a sustainable-income concept. Second, the capacity exists to rectify this omission with respect to the value of forests that is linked to marketed production.

While adjustments in an asset account are conceptually similar to net investment of "made assets," for forests it is more precise to call the change in asset values net accumulation to reflect the fact that, even at constant prices, the asset value of a forest can either increase or decrease. Most generally, net accumulation is defined as the change in an asset

---

[9]USDA Forest Service (1995) also present estimates based on fees collected (which show much lower value overall and relatively less for recreation and wildlife); willingness to pay, including consumer surplus (which show higher overall values and greater importance for recreation and wildlife); and income generated, including that generated by downstream activities such as lodging and equipment rentals related to forestland recreation (which show the highest overall value). From the perspective of comparability with the current national economic accounts, the methods associated with the discussion in the text are preferable to the other three methods.

value from one period to the next. Because asset values cannot generally be inferred, economists infer the value of the asset from assumptions about timber markets. A full analysis of this issue is presented in Appendix C. Three major alternative approaches to accounting for changes in asset values of forests are described below.

Hotelling model. The first approach is analogous to the literature on nonrenewable resources discussed in Chapter 3. In a sense, this approach treats the exploitation of primary, old-growth forests as timber mining. Since it is generally uneconomic to replace primary forests with forests of a similarly old age, this analogy is not as odd as it might appear. Under these circumstances, the change in the value is the volume of the harvest times the difference between the price and the marginal extraction cost. This model of net accumulation is called the Hotelling model to emphasize the connection between mining old growth that *will not* be replaced and mining minerals that *cannot* be replaced.

Based on historical studies, this approach appears to be a reasonable approximation of empirical trends in forest development (see Berck, 1979; Lyon, 1981; Sedjo and Lyon, 1990; and Sedjo, 1990). In the early stages of development, net growth of the forest is nil: photosynthesis just balances the death of plant tissues and entire trees. Because growth is nil, any harvest at all exceeds the growth of the forest. Since the harvest is greater than the growth, the timber inventory declines. As the inventory of old-growth timber declines, timber becomes more scarce, and timber prices rise. In addition, harvesting costs increase as logging extends into increasingly remote sites. Prices rise until the purposeful husbandry of second-growth timber and the use of nonwood substitutes (stone, concrete, and steel for construction; fossil fuels, solar energy, and conservation for energy) becomes economic. This analysis is broadly consistent with the development of the forest sector in the United States. Harvest exceeded growth until the 1950s. Timber prices rose at a real rate of about 4.6 percent per year between 1910 and World War II and 3.1 percent per year from that period to the mid-1980s (Clawson, 1979; Sedjo, 1990; and Binkley and Vincent, 1988).

Transition models. While the Hotelling model may be appropriate for the case of pure depreciation under the assumption of perfect capital markets,[10] it misses several important aspects of the forest sector, includ-

---

[10]The Hotelling model assumes perfect capital markets in which the rate of return in the mining or old-forest sector equals the rate of return in alternative economic activities. In countries, especially developing countries, where both forest and mining activities earn disproportionally high returns because of special favors and licenses, the Hotelling model is not appropriate. It greatly overstates the true decline in the value of these stocks as they are mined.

ing (1) "discovery" of new old-growth forest stocks (e.g., the rapid expansion of logging in the British Columbia interior to serve U.S. markets once U.S. prices had risen to the point that accessing this comparatively remote region became economic), and (2) the fact that the old-growth forests were replaced with faster-growing second-growth forests. Both effects attenuate price increases, causing the ordinary Hotelling model to overstate forest depreciation. These effects are the forest analog of mineral deposits analyzed in Chapter 3.

Transition models account in part for these problems by recognizing that forest growth offsets harvests. Assuming constant prices and a forest inventory recognized only by total net growth, this model suggests net accumulation is given by the difference between price and marginal harvesting cost times *growth minus harvesting* (rather than simply minus harvesting in the Hotelling model). By recognizing forest growth, such a formulation improves on the ordinary Hotelling approach, but still suffers the defects of (1) ignoring endogenous price changes in the sector, and (2) characterizing the forest only by net growth and not its more complex underlying age-class structure.

Managed second-growth forests. Economic theory suggests that, once the transition between old- and second-growth forests is complete, timber prices will stabilize, and the economic return to holding forests will arise solely from forest growth. Vincent (1997) has analyzed this case and developed the appropriate measures of net accumulation for optimally managed second-growth forests. The appropriate estimate of the value of asset accumulation is more complicated here (see Appendix C for a full discussion). Accumulation depends on the forest age structure, discount rate, timber-yield function, and economically optimal rotation age. While this approach improves on both the Hotelling and transition approaches, certain shortcomings remain. In particular, this approach assumes that forest owners cut their trees at the economically optimal time and that timber prices grow at a constant rate. This theory of forest valuation can be used to formulate a practical approach to measuring the economic depreciation of forests. Before turning to that recommended approach, it is useful to examine BEA's work on forests and the international literature in this field.

BEA's Approach and International Comparisons

As noted, forests are part of Phase II of BEA's IEESA effort. As a consequence, BEA's work on forests to date has not been extensive and may need refinement (see Howell, 1996). In its current work, BEA separates forestland from the timber inventory. "Forests and other wooded land" are valued at the average value of agricultural land. In general, edaphic and geomorphologic factors make forestland less valuable than agricultural lands, and the rate of change in forestland prices is uncorrelated with the rate of change in farmland prices (see Washburn, 1990). BEA updated their estimates of the timber inventory each period using separate Forest Service estimates in physical terms of growth and removals. Starting with physical inventory estimates, BEA added physical estimates of growth (additions) and removals (depletion) to derive closing stocks. Each year's closing stock estimate became the following year's opening stocks (except in the Forest Service inventory years, when inventory estimates of standing timber were used). Opening and closing stocks, additions, and depletions were then valued at the stumpage prices; the difference between the opening stocks plus additions less depletion and closing stocks, in monetary terms, was placed in revaluations.

BEA uses the Hotelling model to value the timber stock in each period. Timber is valued at the national average stumpage rate, with species divided into two categories, softwood and hardwood. When measured at a national level, marginal extraction costs are probably nonzero (production increases are accomplished by turning to increasingly costly regions). There is some evidence that extraction costs are constant within regions, however (Adams, 1997). One conceptual flaw in BEA's current approach is that it measures the depreciation of recreational land on the basis of the costs of repair and maintenance of federal government expenditures for parks. The panel has noted in numerous places the flaw in this approach. Having accounted for one of the costs of providing recreational services, BEA does not adjust national income to reflect the benefits. BEA recognizes the criticisms of this approach and plans to use other approaches in the future. BEA publishes a full account for 1987, although it produces data on the value of timber stocks for 1952-1992. Using BEA's data, the net accumulation of timber in 1987 was $2.1 billion at 1987 prices and $47.0 billion if price changes are included.

While BEA's methods can and should be refined as the environmental accounts are developed, they are consistent with current international practice. Table 4-4 provides a summary of 29 studies from around the world that have attempted to extend the treatment of forests in national income and product accounts. Most of these efforts use variants of the so-called "net price" approach (see equations C.3 and C.4 in Appendix C). Many fail to distinguish marginal and average extraction costs. Accounting for net timber accumulation is well established in the international literature. None of the studies appears to use the third method described in the previous subsection of a managed second-growth forest.

TABLE 4-4   Summary of Forest Accounting Studies

Study Area	Reference	Valuation Method			
		Net Price	El Serafy	NPV	Other
Global	World Bank (1997)	T		√	
Asia	Vincent and Castaneda (1996)		G		
Australia I	Young (1993)			√	
Australia II	Skinner (1995), Joisce (1996)	H		√	√
Austria	Sekot et al. (1996)	H		√	√
Canada I	Anielski (1992a, 1992b, 1994, 1996)	T			
Canada II	Statistics Canada (1997), Baumgarten (1996)	H		√	
Chile	Claude and Pizarro (n.d.)	?	?	?	?
China	Li (1993)	T			
Costa Rica I	Repetto et al. (1991)	?	?	?	?
Costa Rica II	Aguirre (1996)	T			
Ecuador	Kellenberg (1995)	T			√
Finland I	Koltolla and Mukkonen (1996)	T			
Finland II	Hoffrén (1996)	T			
Indonesia	Repetto et al. (1989)	T			
Malaysia I	Vincent et al. (1993)	T			
Malaysia II	Vincent (1997), Vincent et al. (1997)		G		
Mexico	van Tongeren et al. (1993)	T	√		
Nepal	Katila (1995)	T			
New Guinea	Bartelmus et al. (1992, 1993), Bartelmus (1994)	X	X	X	X
New Zealand	Bigsby (1995)	H			
Philippines I	IRG et al. (1991, 1992)	T		√	
Philippines II	Cruz and Repetto (1992)	T			
Sweden I	Hulkrantz (1992)	T			
Sweden II	Eliasson (1996)	T			
Tanzania	Peskin (1989a)	X		X	X
Thailand	Sadoff (1993, 1995)	T			√
United States	Howell (1996)	H			
Zimbabwe	Crowards (1996)	T			

Key: H = Hotelling approach; T = transition approach; G = generalized El Serafy approach (elasticity of marginal cost not infinity); X = no timber valuation performed; ? = no information; √ = used technique; NPV = net present value.

Source: Vincent and Hartwick (1997). References in original

A Recommended Approach for
Measuring Net Accumulation of Timber

The three alternative approaches to accounting for changes in asset values of forests discussed above incorporate many restrictive assumptions. The panel investigated other alternatives and identified one (developed by Vincent [1997]) that is similar to the second-growth forests approach, but allows for the possibility that forest managers may deviate from ideal wealth-maximizing behavior. This approach is described in detail in Appendix C. A review of available data indicates that the approach can be readily implemented for the United States using data maintained by the U.S. Forest Service.

Conclusions on Forest Resources

BEA has initiated a useful effort to recognize the economic contributions of forests in the NIPA. Doing so is consistent with a wide international interest in such accounts. The data and methods employed by BEA to date are reasonably consistent with the body of international work in this area. At the same time, data are available for U.S. forestlands that can enable much more complete estimates of net timber accumulation than either those developed to date by BEA or those available in the literature for other countries. BEA could fruitfully work with the U.S. Forest Service in developing annual estimates of net timber accumulation using these data.

This work could also be related to other important values of the forest, particularly recreation and other nonmarket activities. While the data and analytical methods are not yet adequate to provide precise estimates of the value of all forest-sector flows to the economy, nonmarket forest values for the nation as a whole appear to exceed the value of timber by a substantial amount. Many of these forest values (such as recreation or self-produced fuel wood) are best understood conceptually in the context of household production. The household combines specific aspects of the forest resource with household capital and labor to produce valuable nonmarket goods and services. Viewed in this context, forests present many of the same challenges for national accounting as do such important products and services as home-cooked meals and in-home education or childcare. It is therefore logical for BEA to consider these aspects of environmental accounting as part of the larger problem of valuing the contributions of nonmarket activity to economic well-being.

In conclusion, constructing a set of forest accounts is a natural next step in developing integrated economic and environmental accounts. At the same time, it must be recognized that there are many thorny problems involved in forest accounting. Given the available data and methods, the panel concludes that this accounting is a useful next step in developing the IEESA.

AIR QUALITY: A PUBLIC ENVIRONMENTAL GOOD

Air quality is one of the most important examples of a public environmental good and thus should be among the top priorities for inclusion in environmental accounts. It also presents issues for environmental accounting similar to those encountered with other environmental assets, such as water quality and climate change. Severely degraded air quality in many cities of the United States in the 1960s generated a number of federal regulations during the early 1970s designed to reduce emissions of pollutants that contributed to this degradation. Air quality has many dimensions, and early regulations focused on some of the more obvious and easily addressed problems. As scientific research further illuminated the less immediately obvious impacts of degraded air quality, such as chronic effects on health, these earlier controls were tightened, and new regulations addressed a wider range of pollutants.

The first subsection below examines the various market and nonmarket impacts of air quality. The second reviews some major pollutants that result in degradation of air quality and their primary physical effects. This is followed by review of a recent attempt to estimate comprehensively the benefits associated with improvements in air quality. The fourth subsection addresses the relevance of these damage estimates to environmental accounting. The section ends with the panel's conclusions on accounting for air quality.

Air Quality Impacts on Market and Nonmarket Activities

Degraded air quality can have a harmful effect on both market activities (e.g., reduced crop yields or lost work-days) and nonmarket activities (e.g., losses due to illness beyond those related to paid labor, such as those to retired persons, and reduced amenities in recreational facilities). These air quality effects belong in the production accounts of environmental accounts. Moreover, degraded air quality can affect the value of natural-resource assets (e.g., acid deposition damage to forests), can cause deterioration of physical capital (e.g., damage to the exterior of buildings), and has long-term health impacts that affect human capital (e.g., premature death and effects of lead on measured IQ of children). Such effects might be included in the asset component of environmental accounts. With assets as with production, there are both market and nonmarket effects: market impacts include capital asset deterioration and forest timber loss, while nonmarket impacts include lost value due to damaged landmarks or degradation of forests for recreational purposes.

Major Air Pollutants and Their Health and Ecological Effects

Table 4-5 lists some important health and ecological effects of exposure to six air pollutants for which the U.S. Environmental Protection Agency (EPA) has established National Air Quality Standards—carbon

TABLE 4-5   Environmental Protection Agency's Six Criteria Air Pollutants

Pollutant Trends (1986-1995)		Major Effects	Leading Source
Ground-level ozone (O_3)		Respiratory illness/lung damage	Transportation* (37%)
			Solvent utilization (28%)
Concentration	−6%	Crop/forest damage	
Emissions	−9%	Building/material damage	
		Visibility problems	
Carbon monoxide (CO)		Reduced oxygenation of blood	Transportation (81%)
Concentration	−37%	Heart damage	
Emissions	−16%		
Sulfur dioxide (SO_2)		Respiratory illness	Electric utilities (66%)
Concentration	−37%	Building/material damage	
Emissions	−18%	(acid rain)	
		Crop/forest damage	
		Visibility problems	
Nitrogen dioxide (NO_2)		Respiratory illness/lung damage	Transportation (49%)
Concentration	−14%		Electric utilities (29%)
Emissions	−3%	Building/material damage	
		(acid rain)	
		Crop/forest damage	
		Visibility problems	
Lead (Pb)		Infant mortality	Metals processing
Concentration	−78%	Reduced birth weight	(smelters, battery
Emissions	−32%	Childhood IQ loss	plants) (39%)
		Hypertension	Transportation (31%)
		Heart attacks	
Particulate matter (PM-10)		Lung disease	Fugitive dust (68%)
		Mortality	Agriculture and forestry
Concentration	−22%		(20%)
Emissions	−17%		

*Based on volatile organic compounds (VOC) emissions.
Source: U.S. Environmental Protection Agency (1996).

monoxide, ground-level ozone, lead, nitrogen dioxide, particulate matter, and sulfur dioxide. These chemicals are sometimes referred to as "criteria pollutants." In addition, there are many other constituents of the atmosphere that may have impacts of economic consequence. Table 4-6 lists some other components of air pollutants, including air toxins (e.g., benzene), stratospheric ozone depleters (e.g., CFCs), and greenhouse gases

TABLE 4-6   Other Pollutants of Air Quality Identified by
Environmental Protection Agency

Pollutant	Major Effects	Leading Source
Air toxins (188 in total, e.g., dioxins, benzene, arsenic, beryllium, mercury, vinyl chloride)	Thought to cause cancer or other serious health effects, such as birth defects or reproductive effects Ecosystem damage (particularly freshwater fish)	Transportation, wood combustion, chemical plants, oil refineries, aerospace, manufactures, dry cleaners
Stratospheric ozone depleters (e.g., chlorofluorocarbons [CFCs], halons, carbon tetrachloride, methyl chloroform)	Skin cancer Cataracts Suppression of the immune system Ocean food chain stresses	Fossil fuel, industrial cleaners
Greenhouse gases (e.g., carbon dioxide, methane, halogenated fluorocarbons [HFCs])	Broad-scale changes in temperature and precipitation affecting agriculture, health, water resources, recreation, ecosystems Sea level rise	Fossil fuel, combustion, landfills

Source: U.S. Environmental Protection Agency (1996).

(e.g., carbon dioxide and methane). As indicated, EPA has identified 188 air toxins alone.

Exposure to air pollution has a wide range of impacts, including respiratory illnesses (which result from ground-level ozone, sulfur dioxide, nitrogen dioxide, particulate matter, and air toxins); child IQ loss, infant mortality, strokes, and heart attacks (which result from lead); skin cancer (which is the indirect consequence of stratospheric ozone depleters); and increased mortality (resulting from particulate matter, lead, and air toxins) (see Pearce et al., 1996). Ecological effects include impacts on agricultural, forest, and aquatic ecosystems. Airborne chemicals have both positive and negative effects on production of marketed goods and services. Ground-level ozone harms crops, while nitrogen deposition and carbon dioxide enhance plant and timber growth. Ground-level ozone and sulfur dioxide reduce crop yields and timber growth, while air toxins and sulfur dioxide reduce freshwater fish yields. In other cases, atmospheric trace gases have subtle effects that will occur far in the future affecting biological diversity (for greenhouse gases) or ocean food web stresses, and ultimately causing severe sight damage for many mammals (for stratospheric ozone depleters).

Table 4-5 also shows the change in emissions and sampled concentrations of EPA's six criteria pollutants from 1986 to 1995.[11] Primarily as a result of the Clean Air Act and the Clean Air Act Amendments, emissions of the six primary pollutants have decreased substantially. For example, installing scrubbers and switching to low-sulfur coal caused a 19 percent decline in emissions from coal utility plants, which in turn resulted in an overall 18 percent decline in sulfur dioxide emissions from 1986 to 1995. A 16 percent decline in carbon monoxide emissions during the same period resulted primarily from a 20 percent decline in carbon monoxide emissions of on-road motor vehicles. Similarly, a 32 percent decline in lead emissions was primarily a result of the ban on leaded gasoline.

Declines in nitrogen dioxide (14 percent) and ground-level ozone emissions (6 percent) were less dramatic, but are expected to become more pronounced as the Clean Air Act Amendments of 1990 become effective. For example, reformulated fuel requirements (for oxygen and volatility) for on-road vehicles are likely to reduce carbon monoxide and ground-level ozone emissions. Similarly, the Acid Rain Program (Title IV) requires a 40 percent reduction in sulfur dioxide and a 10 percent reduction in nitrogen dioxide emissions from 1980 to 2010. Particulate matter may be more difficult to control given that almost 70 percent of anthropogenic-related emissions result from fugitive dust (e.g., unpaved roads), with an additional 20 percent coming from agriculture and forestry.

The declines in emissions are, of course, linked to lower concentrations of the six primary pollutants. Whereas emissions are estimated on the basis of industrial activity, technology, fuel consumption, and vehicle miles traveled, concentrations of pollutants are measured at selected monitoring sites across the country. Based on these measurements, estimated airborne concentrations of lead have fallen by 78 percent since 1986, while concentrations of airborne carbon monoxide, sulfur dioxide, and particulate matter have fallen by 37, 37, and 22 percent, respectively. Smaller declines occurred for ground-level ozone and nitrogen dioxide (6 and 14 percent, respectively).

Data on other air chemicals vary widely. Excellent data are available on emissions and concentrations of many of the greenhouse gases (particularly carbon dioxide) and stratospheric ozone destroyers. EPA pres-

---

[11]Data prior to 1986 exist, but cannot be directly compared with data collected from 1986 on because of changes in data collection (see U.S. Environmental Protection Agency, 1996, for more details).

ently monitors national ambient concentrations for few of the 188 air toxins identified in the Clean Air Act Amendments. Rather, the agency sets technology-based performance standards to control emissions of these substances. As a result, EPA has only begun developing a National Toxins Inventory.

Monetized Benefits of Clean Air Regulations

Although a great deal of work has been done on valuing components of air quality, there is currently no comprehensive measure of the economic impacts of air pollution for the United States. However, a recent EPA study evaluating the economic costs and benefits of clean air regulations provides a useful benchmark that sheds light on this issue (U.S. Environmental Protection Agency, 1997). The estimates given are subject to many uncertainties due to the difficulty of estimating exposure and the incidence of effects related to exposure and valuing the effects. In addition, data on air toxins have only recently become available, making it difficult to develop comparable estimates for these pollutants. The EPA study includes no physical or monetary assessments of the impacts of changes in air quality on ecosystem health, physical capital, or global public goods, such as slowing climate change and preventing ozone depletion. Moreover, many of the estimates of benefits, particularly those involving the valuation of health benefits and the discount rate, have been the subject of major criticism (see Clean Air Act Council on Compliance, 1997).

Notwithstanding these limitations, the EPA study provides an indication of the overall economic importance of changes in air quality, as well as a sense of the relative importance of the various air pollutants and the impacts on different sectors. The study estimates the economic benefit of actual air pollution relative to a counterfactual baseline that assumes no controls imposed after 1970; roughly speaking, the counterfactual is for emissions to grow with the economy, rather than declining as described above. The major result presented is that the economic benefits of reduced air pollution in 1990 are estimated to be worth $1,248 billion. Reduced mortality benefits ($1,004 billion) account for 80 percent of this total; together, avoided human health effects account for 99 percent of the total. In addition, benefits of improved visibility are estimated at $3.4 billion, those of reduced household soiling at $4.0 billion, and those of increased agricultural income from reduced yield losses due to ozone at about $1.0 billion. With regard to specific pollutants, most of the benefits are attributed to reductions in particulate matter (PM-10) and lead; the benefits of ozone reduction are estimated to be only on the order of $2 billion.

Caution is warranted in drawing too many conclusions from these estimates and comparisons. Certain assumptions might have had the effect of exaggerating the economic benefits, and there are major uncertainties about the health impacts, particularly because of weaknesses in human exposure data. Moreover, the study omits some of the major effects of acid deposition on forests, lakes, and buildings, and the impact of tropospheric ozone on ecosystems is not valued. The figures presented should therefore be viewed as order-of-magnitude estimates. Even with all these qualifications, however, it appears that the economic impacts of air quality on human health are highly significant.

Air Quality Benefits and Supplemental Accounts

The estimates of the benefits of pollution control just discussed reflect the value of changes in the level of air pollutants resulting from proposed regulations. They are relevant for regulatory or cost-benefit purposes, but they are not the appropriate values for economic accounts. Production accounts should measure the damages associated with remaining levels of pollution, in terms of both production accounts and change in asset values. This difference between abatement and residual damage can be quantitatively large. For example, ozone concentrations fell only 6 percent between 1986 and 1995. As a result, regardless of the benefits of preventing higher levels of ozone than those of 1986, the value of changes in ozone concentrations over this period would be relatively small. In contrast, lead and PM-10 concentrations fell 78 and 22 percent, respectively, over the same period, and consequently the damages from these chemicals would be much smaller in 1995 than in 1986. In other words, whereas comprehensive consumption would have a substantial negative entry due to lead and PM-10 in 1986, the negative values would be of much smaller magnitude in 1995. The result might be a substantial increase in the estimate of growth of comprehensive consumption over this period.

As discussed earlier, air pollution affects production activities, assets, and nonmarket activities. Most of the estimates from the EPA study refer to the production accounts: days of work lost, shortness of breath and acute bronchitis, loss of visibility, and crop losses are effects on production activities. Crop losses and the output losses from lost work-days are already included implicitly in the accounts because these relate to market activities. Supplemental accounts that would identify these losses separately would serve to connect them specifically to air pollution. The estimates for shortness of breath and acute bronchitis include both damages that may already be reflected in the production accounts (i.e., reduced worker productivity while on the job) and damages that would be reflected only if the accounts were expanded to include household production (e.g., impacts on tennis and jogging). Many of the effects not estimated by EPA, such as those of acid deposition on forest health, freshwater quality, or ecosystem function, would also include effects on both market activities already in the accounts, such as timber or commercial fishing, and nonmarket goods, such as recreation.

Asset effects present greater complexity, as was seen above for the case of forests. Some impacts, such as those on soil or fish farms, would be reflected in the market value of these assets. Others, such as mortality and chronic bronchitis, are long-term effects on human resources. These effects would require adjustments in the asset accounts if a full set of asset accounts for human health and capital were constructed.

One particular concern arises if the accounts are to include the impact of air pollution on human health. The impact of air pollution and other environmental activities on human health is often taken out of the context of other health-related activities. If one were to track environmental trends alone, it might be concluded that until the 1970s, growing environmental problems were leading to a deterioration in the health status of Americans. This conclusion is, in fact, incorrect. Activities outside the environmental arena—including improved sanitation, vaccinations, and public-health measures—led to improved life expectancy over the first seven decades of this century. It would therefore be misleading to enter only a large health negative into a set of augmented income accounts. The positives and negatives in the environmental entry in a set of health accounts would have to be placed in the context of the vast changes in health status of the American population.

Conclusions on Air Quality

The basic finding emerging from the above discussion is that air quality is likely to be a major nonmarket effect. While EPA's estimates of benefits of $1.2 trillion per year due to reduced air pollution are highly uncertain, do not include all effects, and measure a somewhat different concept than would be appropriate for the accounts, it is likely that a realistic assessment of reduced damages due to improved air quality would yield a much larger figure than the $27.1 billion in air pollution control expenditures used by BEA as a placeholder. In the panel's view, no other area of natural-resource and environmental accounting would have as great an impact as the potential correction from air quality. The magnitude of this impact indicates that the development of supplemental accounts for air quality is a high priority. Indeed, the overall review of augmented accounting in Chapter 2 reveals only a few areas close in importance, such as the value of leisure, health status, and nonmarket educational investments.

At the same time, air quality is a most elusive concept since it has so many different components. To include these effects in the accounts, several data and measurement obstacles must be overcome. First, determination of the physical impacts of changes in air quality, generally estimated through dose-response functions, should be focused on the effects of actual human exposure to air pollution. Second, the damage estimates must separate the market effects of changes in air quality that are currently captured in the accounts (lost productivity) from the nonmarket effects that are not currently captured (lost leisure activities). Third, there is a need for reliable and objective physical and monetary damage estimates associated with exposure to air pollutants, including air toxins, ozone depleters, and greenhouse gases. Fourth, significant data gaps with respect to the impacts of air pollution and changes in air quality on ecosystem health must be filled. And finally, the estimates must represent year-to-year changes, rather than changes from a hypothetical level of pollution without regulations.

Developing a set of accounts in this area, along with the associated physical measures and valuations to apply to those measures, is a major long-run task for the nation. This task far transcends the scope and budget of BEA, and much of the necessary work lies outside BEA's specialized expertise. The task for the short run, therefore, is to continue basic research on the underlying science and economics of estimating the benefits of public goods such as clean air. Many years of concerted research are likely to be required before the materials for a set of augmented accounts in this area are available. But the payoff from the research would be large, not only in producing the raw materials for improved environmental accounts, but more important in providing the data and analysis needed for improved public policy concerning the environment. In short, the task of constructing environmental accounts for important public goods should be part of a more general goal of improving the nation's information and analytical systems in this area.

CONCLUSIONS AND RECOMMENDATIONS ON RENEWABLE AND ENVIRONMENTAL RESOURCES

General Approach

4.1 The panel recommends that BEA continue its work toward accounting for changes in natural-resource assets and for the flow of services from these assets.

Environmental variables affect economic well-being in three major ways: direct effects on consumption or income of households, industry, and government; accumulation in the environment of stocks of residuals that then affect economic activities or economic assets; and effects on the service flows of economic assets, including capital stock, natural resources, and human resources. The main value of natural-resource accounting is in providing a complete picture of the role these resources play in the economy. Sometimes this information can be used to judge the overall sustainability of the use of resources, while at other times it can be used to manage natural and environmental resources and to inform public policy choices.

Valuation

4.2 For valuation, the panel recommends that BEA rely primarily on market values or proxies of market values that are based on actual behavior. Contingent valuation, while sometimes useful for other purposes, is currently of limited value for environmental accounting in the context of the economic accounts.

Valuing environmental goods and services requires distinguishing between private and public goods. Market prices provide the marginal valuations for private goods, but determining the value of public goods requires the summation of individual values. Moreover, there may be no behavioral traces for individual valuation of public goods.

Price data are relatively reliable for private market goods produced from forest and agricultural assets, such as timber stumpage, livestock, and land use and quality. Values for near-market goods—those that have direct counterparts in the market—can be constructed by comparing the near-market goods with their market counterparts, adjusting for quality as necessary. Techniques for valuation of public goods are still under development. Some techniques—such as hedonic or travel-cost studies—rely on behavioral or market-based estimates; while these estimates are subject to significant measurement errors, they are conceptually appropriate in economic accounts. Other techniques, such as contingent valuation, are not based on actual behavior, are highly controversial, and are subject to potential response errors.

Quantitative Data

4.3 Quantitative data on many natural-resource assets are currently relatively adequate. However, the data on many environmental variables are at present poorly designed for the construction of environmental accounts. The panel recommends that greater emphasis be

placed on measuring effects as directly as possible. Of particular importance are measures of actual human exposure to air and water pollutants, rather than modeled measures of exposure based on ambient pollutant levels at current monitoring sites.

Quantitative data for natural resources are often of high quality relative to the other quantitative data in the NIPA because there are well-established units of measure for many natural resources. Quantitative data on near-market activities such as fuel wood for own use are conceptually straightforward, and many of these data are currently collected by federal agencies. Measurement of nonmarket goods and services and explicit accounting for quality changes, particularly for those that have public-good characteristics, are currently subject to severe methodological difficulties and insufficient data. There are relatively good data on emissions of many residuals from industrial and human activities, but for most harmful pollutants except lead there is very little systematic monitoring of human exposures.

Inclusion of Public Goods

4.4 The panel finds that more work will be needed on techniques for establishing production flows and values for the assets and services of public goods to place them on a comparable basis with the prices and quantities used in the core accounts.

True public goods, for example biodiversity, species preservation, and national treasures such as the Florida Everglades and Yellowstone National Park, present severe conceptual and measurement issues for incorporation into a national accounting system.

Data Collection

4.5 The panel encourages BEA to help mount a concerted federal effort to identify the data needed for measuring changes in the quantity and quality of natural-resource and environmental assets and associated nonmarket service flows.

Many different federal agencies collect data or have expertise that will be essential to BEA, particularly as its efforts expand to include Phase III assets and associated flows. BEA already cooperates with other agencies in collecting data for the core accounts; supplemental environmental accounts will require cooperation with, for example, the Environmental Protection Agency, the Department of Agriculture, the Department of the Interior, the Bureau of Labor Statistics, the Bureau of the Census, the Energy Information Administration, the National Institute of Environmental Health Sciences, and the Department of Health and Human Services.

Regional Resolution

4.6 The panel recommends BEA focus on developing supplemental accounts for the nation as a whole as a first priority. At the same time, BEA should preserve regional detail where it exists so that these data are available for analysts interested in developing accounts at the regional level.

The development of national estimates will require sampling, measurement, and valuation techniques that reflect the fact that the quality and value of natural-resource assets and associated flows vary geographically. While some assets and flows may not be important to the national economy, they could be far more important to regional and local economies.

Next Steps

4.7 The panel recommends that funds be provided to reinitiate and improve the design of the collection of data on pollution control and abatement expenditures.

4.8 As BEA further develops its natural-resource and environmental accounts, an important step is to incorporate near-market goods and services—those that have close counterparts in marketed goods and services. There is a clear basis here for measuring quantities and establishing values in a manner comparable to that used for the core accounts.

4.9 Construction of a set of forest accounts is a natural step in developing integrated economic-environmental accounts. The United States has much of the data needed for such an effort, and the analytical techniques are relatively well developed.

4.10 Based on available information, the economic impacts of air quality are likely to be the most significant element in the environmental accounts; development of such accounts is a central task for environmental accounting. At the same time, because of the unresolved conceptual issues and the need for appropriate physical measures, the development of stock and flow accounts for air quality and other important public goods poses awesome difficulties. This task far transcends the scope, budget, and expertise of BEA. A major goal for the near term is to continue basic research on the underlying science and economics in this area.

5

Overall Appraisal of Environmental Accounting in the United States

 This chapter contains the panel's overall conclusions and recommendations, which are based on the analysis and findings presented in previous chapters; specific conclusions and recommendations related to accounting for subsoil mineral resources and for renewable and environmental resources are presented in Chapters 3 and 4, respectively. The sections that follow address in turn the basic questions that arise in constructing integrated environmental and economic satellite accounts, the budgetary implications of developing environmental accounts, and issues of data and implementation.

FUTURE DIRECTIONS FOR THE U.S. INTEGRATED ENVIRONMENTAL AND ECONOMIC ACCOUNTS

This section presents the panel's overall conclusions and recommendations with regard to eight key questions related to the construction of integrated environmental and economic accounts:

1. What is the role of natural-resource and environmental accounting?
2. What is the value of augmented nonmarket accounts?
3. Should the Bureau of Economic Analysis (BEA) resume work on the Integrated Environmental and Economic Satellite Accounts (IEESA)?

4. Should the United States pursue a phased or comprehensive approach to augmented national accounts?
5. Should the IEESA be developed in the core or satellite accounts?
6. What is the relationship of the IEESA to the United Nations System of Environmental and Economic Accounts (SEEA)?
7. What are appropriate techniques for measuring quantities and values for nonmarket activities in the national accounts?
8. What should be the next steps in extending the IEESA?

1. What Is the Role of Natural-Resource and Environmental Accounting?

BEA has developed integrated environmental and economic accounting in response to Presidential directives, as well as the growing interest in and importance of the subject (see Bureau of Economic Analysis, 1994a). Work on environmental accounting has been conducted over the last quarter-century under several administrations. Environmental accounting was introduced during the Ford Administration, when Secretary of Commerce Elliott Richardson called for environmental accounting to track capital investment expenditures on pollution abatement. This initiative was further developed by the Carter Administration. In 1990, the Council of Economic Advisers under President Bush recommended that BEA expand its work on environment-economy interactions. And in 1993, BEA was given a mandate by the Clinton Administration to develop first-phase resource accounts within the framework of the national accounts and to pursue construction of the IEESA.

Natural-resource and environmental accounting has been studied extensively by the United Nations and the European Union and is currently an area of intensive research in all major countries.[1] Many countries have developed additional accounts for minerals, forests, and pollution-control expenditures. The broad-based research that has been conducted on environmental accounting is an indication of the high priority assigned to the development of integrated environmental and economic accounting in the United States and other countries.

As discussed further below, better natural-resource and environmental accounts would provide valuable insights into the interaction between the environment and the economy. They would also provide information

---

[1]The Organization for Economic Cooperation and Development (OECD) Council of Environment Ministers, the United Nations Conference on Environment and Development, the heads of government of the Group of Seven, the "London Group" of National Income Accountants, and numerous other international bodies have recommended that nations develop integrated environmental and economic accounts.

on the implications of public and private investment and consumption decisions, and help determine whether the nation is running down its stocks of natural resources and environmental assets in an unsustainable manner. Better accounts can inform the nation about the implications of different regulations, taxes, and consumption patterns and thereby lead to more efficient economic, environmental, and natural-resource policies.

There is also a close connection between current approaches to augmented income and product accounts and measures of sustainable income. As discussed in Chapter 2, properly constructed national income and output can be interpreted as the maximum sustainable per capita consumption. Ideal measures of sustainable income include all consumption items (including the values of nonmarket consumption), along with the value of changes in the stocks of different assets. These ideal measures of national output and sustainable income can serve as a useful guide to the United States as it improves its national accounts by extending their boundaries.

5.1 The panel concludes that extending the National Income and Product Accounts (NIPA) to include assets and production activities associated with natural resources and the environment is an important goal for the United States. Environmental and natural-resource accounts would provide useful data on resource trends and help governments, businesses, and individuals better plan their economic activities and investments. The rationale for augmented accounts is solidly grounded in mainstream economic analysis. BEA's activities in developing environmental accounts (IEESA) are consistent with an extensive domestic and international effort to both improve and extend the NIPA.

2. What Is the Value of Augmented Nonmarket Accounts?

Developing natural-resource, environmental, and other nonmarket accounts is an investment in better information for the nation. Well-designed environmental accounts can overcome the recognized shortcomings of the current market-based accounts and provide information about the interaction between the economy and the environment that would support private and public decisions. There are three principal reasons why developing a set of environmental and nonmarket accounts would benefit the nation.

First, comprehensive accounts give a complete picture of economic activity; by contrast, traditional national accounts, which cover only market transactions, provide a misleading indicator of economic activity. Comprehensive accounts contribute to a better understanding of the functioning of the economy and of the interaction between the economy and the natural environment. Businesses and governments need and want to know about basic market conditions in the world, the nation, and their region. Without good market and nonmarket information, firms are flying blind.

There are many examples of how conventional accounts send misleading signals about economic activity. When companies discover large deposits of oil, gold, and other mineral assets, these are not counted in the nation's investments or as increases in its wealth. Similarly, even though forests contribute greatly to the nation's well-being, only timber production is counted in the national output. The value of hunting, fishing, and other forms of nonmarket forest recreation is not counted as part of the national output even though the total economic contribution of these nonmarket forest outputs probably exceeds the value of the timber production (see Chapter 4). Outside the environmental sector, traditional accounts provide misleading estimates of economic activity because they omit nonmarket production and investment in important areas such as human capital and education and nonmarket work at home.

The largest distortion in the environmental area probably arises in the sectors relating to environmental quality. Economic studies reviewed in Chapter 4 indicate that the nation is devoting more than $100 billion annually to pollution abatement and control expenditures. Yet many of the economic benefits derived from these expenditures are omitted from the national accounts. Even though investments in clear air and water produce benefits in improved health of the population, improved functioning of ecosystems, improved recreational opportunities, and lower property damages, virtually none of these benefits are captured by current market-based economic accounts.

Second, environmental accounts would provide important information for management of the nation's public and private assets and for improved regulatory decisions. For example, enhanced natural-resource and environmental accounts can provide useful information on natural assets under federal management. Better information on the value of minerals on federal lands would be useful in determining appropriate royalty rates and leasing policies for resources not allocated through competitive auctions. For renewable resources, better information on the stumpage value of timber in national forests would be useful not only for accounting purposes, but also for improved management of these forests and for decision making on the balance of different uses among timber harvesting, wilderness preservation, recreation, and other uses. Better information on fisheries would be valuable to federal agencies responsible for management of these assets.

In the case of environmental resources such as air and water quality, a comprehensive set of environmental accounts would provide useful information on the economic returns the nation is reaping from its environmental investments. The contrast between private and public investments is instructive in this regard. When a private company invests in an automobile factory or a power plant, company accounts can be used to estimate the economic costs and benefits of that investment. In contrast, even though the nation has allocated more than $1 trillion to environmental, health, and safety investments over the last quarter-century, it has no accounts by which to reckon the returns to those investments. Improved environmental accounts would also provide essential information for sound benefit-cost analyses in regulatory decision making. One of the most serious weaknesses in the U.S. environmental database is the lack of comprehensive and reliable data on actual human exposures to major pollutants. Better information on physical emission trends, human exposures, and the economic impacts and damages due to air and water pollution would be valuable for expanded accounting measures of productivity. Hence, both the underlying information and the aggregate dollar estimates in environmental accounts would provide valuable information for ensuring that the nation's environmental regulations pass an appropriate cost-benefit test.

Third, investing in improved accounts would have a high economic return for the nation. The federal government currently invests substantial amounts in collecting, analyzing, and distributing statistical data on the nation. Provision of statistical data is an investment because information is a public good. The gathering of high-quality, comprehensive, and timely data on economic activity requires the resources and data-collection abilities of the government. But the federal government has to date invested very little in the development of nonmarket economic accounts. And while many in the private sector have attempted to construct such accounts, private researchers have neither the resources nor the data required to do so. As a result, the United States today has no set of comprehensive economic accounts, public or private.

There are many examples of the economic benefits of comprehensive economic accounts. One area in which environmental data have proven valuable is analysis of the relationship between environmental regulation and productivity. A second area involves improving understanding of the costs and benefits of environmental regulations. Existing data and studies do not provide sufficient detail to allow pollutant-by-pollutant or sector-by-sector estimates of costs and benefits. Improved accounting systems for the environment can help sharpen both estimates and regulatory tools so that pollution control investments can be more effectively allocated. Yet a further important application with substantial potential value for the nation is management of our public lands.

An area of growing importance is analysis of the economic costs and benefits of steps to slow greenhouse warming. The United States is considering a major commitment to reduce its greenhouse gas emissions. Better estimates of the sources and sinks of these gases, particularly in forests, could help reduce the costs of meeting this commitment. This area represents one of the most dramatic examples of the benefits of establishing comprehensive nonmarket physical and economic accounts, involving potential savings to the nation in the tens of billions of dollars annually.

5.2 The panel concludes that developing a set of comprehensive nonmarket economic accounts is a high priority for the nation. Comprehensive accounts would address such concerns as environmental impacts, the value of nonmarket natural resources, the value of unpaid work, the value of investments in human capital, and the uses of people's time. A set of comprehensive accounts would illuminate a wide variety of issues concerning the economic state of the nation.

3. Should BEA Resume Work on the Integrated Environmental and Economic Satellite Accounts (IEESA)?

The central issues discussed in this report are whether BEA's IEESA represent a useful activity for the United States and whether work on the IEESA should resume. In addressing these issues, the panel is concerned that, particularly since the congressional stop-work order of 1994, the United States has fallen behind in developing environmental and other augmented accounting systems. The United States has in place today only the bare outline of a set of extended environmental accounts, with numerical estimates limited to subsoil mineral assets; the nation has no set of satellite environmental accounts, no physical accounting system, and no environmental input-output system.[2]

In weighing future directions for environmental accounting in the United States, the panel offers three general conclusions, which are followed by three associated recommendations. First, it is clear that there are many alternative approaches to natural-resource and environmental accounting. Given BEA's expertise, along with its limited resources, BEA's phased approach is a reasonable alternative. As noted earlier,

---

[2]The Netherlands and Denmark have done considerable work on the requirements and construction of an environmental input-output system. This work would be useful in understanding the data requirements for an input-output system for the United States. Fostering the development of such data will be an impetus for developing input-output models. See de Boo et al. (1991) and Jensen and Pedersen (1998).

however, the shortcoming of the phased approach is that it is looking only where the lights are brightest and not where the needs are greatest. It is important, therefore, for the United States to develop the accounts in areas not illuminated by the bright light of market transactions. Developing a comprehensive set of nonmarket accounts is the most promising alternative to such a limited focus. In a country of the size, diversity, complexity, and wealth of the United States, providing this information is an essential function of government and one the federal government is supporting insufficiently at present.

Second, the task of developing a comprehensive set of nonmarket accounts for the United States is a large undertaking that would stretch the scope and specialized expertise of BEA. Moreover, if undertaken within the resources currently projected, such a task would clearly result in cutting back other important functions and proposed improvements planned by BEA. The panel therefore cautions that any serious attempt to develop environmental accounts will require additional funding. One potential approach, discussed in detail in the final section of this chapter, would be for BEA to undertake this project jointly with other agencies that are oriented to natural-resource and environmental issues. These agencies have considerable expertise in the analysis of environmental and nonmarket activities and would be useful partners in providing the data and developing prototype systems for nonmarket accounts.

Third, the panel is mindful of BEA's important mission and of the precious nature of the data on marketed economic activity it provides. In addition to providing key macroeconomic data and information on different sectors of the economy, BEA has been highly innovative in introducing new approaches, such as improved price and output indexes, and in enhancing the quality of its data on services and international transactions. These data cannot be provided by the private sector and are an important public good. The panel therefore emphasizes that appropriate support for these core activities of BEA is of paramount importance. Activities to develop environmental accounts should be incremental to ongoing activities and improvements and should not come at the expense of core activities. We recommend below that support not be at the expense of BEA's core activities. It is also important that the relevant work of other agencies in supporting these activities (such as the Bureau of the Census, the Bureau of Labor Statistics, the Environmental Protection Agency, and the U.S. Department of Agriculture) be adequately supported.

5.3a The panel was charged to analyze BEA's initial effort in constructing its environmental accounts. Having reviewed existing studies by BEA and other U.S. agencies, by other national statistical agencies,

by international agencies, and by private researchers, the panel concludes that BEA should be commended for its initial efforts in developing a prototype set of environmental accounts for the United States. With very limited resources, it has prepared a set of useful subsoil mineral accounts. BEA's methodology is based on widely used and generally accepted principles, and the agency has relied on sound and objective measures in developing these prototype accounts.

5.3b Developing a full set of natural-resource and environmental accounts would contribute significantly to understanding of the interactions between economic activity and the environment in the United States. Improved accounts would allow a better understanding of productivity, sustainability, and the environment; they would facilitate better forecasting of future trends and allow the nation to plan for potential critical shortages or environmental problems; and they would enable better public and private decisions on managing the nation's resources.

5.3c Congress should authorize and fund BEA to recommence its work on IEESA development. At the same time, appropriate support for BEA's core activities is of paramount importance to the United States. Activities to develop environmental accounts should be incremental to ongoing activities and improvements and should not come at the expense of the agency's core activities.

4. Should the United States Pursue a Phased or Comprehensive Approach to Augmented National Accounts?

There are two major approaches to developing nonmarket and environmental accounts: a phased approach and a comprehensive approach.

BEA's proposal for the IEESA envisions a *phased extension* of the accounts. The work plan involves developing environmental accounts in three phases. Phase I, completed in April 1994, focused on subsoil mineral assets. The proposal for Phase II is to extend the boundary of the accounts to renewable resources such as timber, fish, and water. Phase III would extend the boundaries to environmental areas such as clear air and water and recreational assets. The new accounts were to be published in supplementary or satellite accounts and would not, in the near future, affect the core NIPA.

In the initial stages, the interactions covered under BEA's plan are those that can be linked to market activities and therefore valued at market prices or at proxies for market prices. This was the rationale for dividing the work plan into the three phases—beginning with subsoil minerals that are entirely within the market economy and proceeding next to renewable resources, such as forests, that are substantially in the market sector. Only after completing its market and near-market accounts would BEA develop accounts for nonmarket environmental resources, such as air and water, and other important nonmarket economic activities, such as education and household work.

An alternative to the proposed BEA work plan is a *comprehensive approach* that would involve developing a broad set of nonmarket accounts in parallel with the near-market accounts. Under this approach, BEA would endeavor to develop accounts not only for the minerals and near-market sectors, but also for nonmarket activities and products, and for environmental and nonenvironmental products and activities.

The panel understands the rationale behind BEA's phased approach to extending the national economic accounts. The advantage of the phased approach is that the effort can draw on the work of other official statistical agencies and researchers and utilize the specialized competence of the agency. The panel is concerned, however, that the phased approach is focused where the light is bright but the terrain is relatively uninteresting—that the narrow focus of the phased approach will limit its usefulness. To reap the full benefit of augmented accounts, it will be necessary to develop nonmarket accounts fully and quickly.

The panel does not underestimate the challenges involved in developing comprehensive accounts that include nonmarket activities. This research is in its infancy, and most of the empirical studies on this topic for the United States have been conducted by private scholars. If the United States is to make significant progress in developing a comprehensive set of nonmarket economic accounts, this work must be undertaken by the federal government under the lead of an established statistical agency such as BEA.

5.4 The panel recommends that BEA develop a comprehensive set of market and nonmarket environmental and nonenvironmental accounts. The panel understands the rationale for BEA's plan to move in phases by first improving its accounts for subsoil mineral assets and then including other market and near-market resources. These steps would provide valuable information for the nation. But the comprehensive approach recommended by the panel would provide more complete, more meaningful, and more useful economic information.

5. Should the IEESA Be Developed in the Core or Satellite Accounts?

At present, BEA does not plan to redefine the core NIPA to include flows or investments in natural resources and the environment. The natural-resource and environmental flows would be recorded in satellite or supplemental accounts. According to BEA, the advantage of satellite accounts is that they provide expanded detail and allow for the exploration of alternative methodologies without reducing the utility of the core national accounts for macroeconomic policy and analysis.

Placing environmental and nonmarket activities in a satellite account implies that these activities would not change the core estimates of gross domestic product (GDP), national income, consumption, or investment. One important reason for placing the IEESA estimates in satellite accounts is to preserve the continuity of the core NIPA, which are an essential tool for assessing the state of the economy and conducting macroeconomic stabilization policy. For example, economic research has shown a close link between movements in GDP and changes in the unemployment rate, changes in tax revenues, and the federal budget deficit. Understanding the economy requires comparing current trends and movements with historical periods in order to forecast the future. To the extent that the national product accounts become incomparable over time, the task of forecasters and policy makers becomes more difficult.[3]

Environmental satellite accounts serve the basic functions of a national accounting system: they provide the raw material needed for policy makers, businesses, and citizens to track important trends and determine the economic importance of changes in environmental variables. One important question is the extent to which depletion of mineral resources is reducing the nation's wealth in an imprudent manner (see Chapter 3). This kind of question can be addressed using the current IEESA mineral accounts for 1987 (as of this writing, later data are not available). In that year, the total change in proved subsoil assets (excluding revaluations) was somewhere between $-0.1 and +3.0 billion (see Bureau of Economic Analysis, 1994a). This figure can be compared with a net investment of $298 billion in "made assets" (which include structures, producer equipment, and inventories, but exclude a wide variety of intangible and other investments, such as those in research and development, software, or human capital). Under the framework of sustainable income developed in Chapter 2, these numbers suggest that the level of investment or disinvestment in subsoil assets was very small relative to the net investment in made assets or capital. The impact of net investment or disinvestment in other natural-resource and environmental assets is likely to be much larger.

Two important issues arise in this context: the appropriate boundary

---

[3]These points are forcefully argued by Okun (1971).

for the core accounts and the state of the art in resource and environmental accounting. One of the fundamental principles of current national accounting is that national income and product occur chiefly within the boundary of the market economy. This boundary is drawn both for practical purposes of data availability and objectivity and because national output is a measure of production of market goods and services. It is also recognized by national accountants that because the core accounts are limited to market transactions, they will not necessarily reflect genuine economic welfare and may provide misleading measures of economic activity and distorted indexes for comparison over time and space (see Chapter 2). Because of the importance of the core accounts for many purposes, it is essential that comparable measures be retained. The core national accounts do not now include, nor would the panel recommend including, nonmarket activities by redrawing the boundary to incorporate, for example, all unwaged work or all natural-resource and environmental activities.

A particularly valuable approach is to present a wide variety of different measures and concepts so policy makers and private-sector analysts can develop their own preferred blend of concepts and measures. The core accounts would, in this view, retain their solid anchor in market transactions, but a wide variety of alternative approaches could be presented as the data and methodologies were developed, reported, and used.

5.5 The panel recommends that the core income and product accounts continue to reflect chiefly market activity. Given the current state of knowledge and the preliminary nature of the data and methodologies involved—especially in those areas related to nonmarket activities—developing satellite or supplemental environmental and natural-resource accounts is a prudent and appropriate decision.

6. What Is the Relationship of the IEESA to the United Nations System of Environmental and Economic Accounts (SEEA)?

Although BEA's proposal for the IEESA is broadly consistent with other international environmental accounting systems, it differs from the SEEA and other systems in some important respects (see Chapter 2). One important conceptual difference lies in the treatment of resource discoveries. Under the IEESA, in contrast with the SEEA, discoveries of resources, such as the proving of oil or gas reserves, are assumed to represent gross investment and therefore to increase both gross and net product measures. There are also some semantic differences in categorization: proved reserves in the IEESA are classified along with other developed assets, while they are treated as nonproduced assets in the SEEA. In addition, soils are classified separately in the SEEA, while in the IEESA they are classified along with agricultural land. A final difference is that environmental degradation in the SEEA is valued at restoration cost and subtracted from gross income along with resource depletion. There is no comparable subtraction with the IEESA, apparently because of an assumption that pollution abatement outlays exactly offset any degradation.

The panel's assessment of these differences is twofold. First, the panel emphasizes that environmental accounting is still an emerging discipline. For this reason, as noted above, it is useful to provide ample room for alternative approaches and experimentation. It would be a mistake to close off promising, untested approaches because they currently appear to have shortcomings relative to other approaches.

Having said this, the panel recommends that in developing its environmental accounts BEA avoid many of the analytically defective shortcuts incorporated in some current proposals. The panel notes that many of the innovations introduced by BEA in the IEESA have a sound economic foundation. For example, the symmetrical treatment of additions and depletions in the minerals account is an economically sound modification of the treatment proposed by the SEEA. However, there is an inconsistency in the current IEESA, which neglect the production-account services provided by environmental assets while including the depreciation of those assets in the asset accounts. This would be analogous in the conventional accounts to including the depreciation of airplanes, but excluding the output or value added of air travel. In this respect, both the SEEA and IEESA appear to equate the terms "nonmarket" and "noneconomic." Omission of the economic services provided by environmental assets conflicts with the objective of permitting better analyses of environmental-economic interactions. Clearly, this conflict can be resolved only as a full set of nonmarket accounts is developed.

Regardless of the eventual direction taken by the U.S. environmental accounts, they should avoid some of the fundamental economic errors characteristic of the IEESA and many other environmental systems. Costs of pollution abatement should not be confused with the benefits of abatement or with pollution damage; depletion is not the same thing as true economic depreciation; and environmental control outlays in a given year never exactly offset environmental damage in that year. Undoubtedly, some of these errors are oversimplifications that were introduced for practical reasons: costs are easier to estimate than damages, depletion is easier to estimate than depreciation, and measuring the actual success of environmental outlays is very difficult. However, there is a real danger that continued uncritical use of such inappropriate proxies will lead to an equivalent uncritical acceptance of their widespread use in environmental accounting systems.

5.6 The panel endorses BEA's development of a set of accounts that are consistent with sound economic principles. In some respects, the IEESA represent a conceptual improvement over the principles underlying the SEEA. Experimentation and diversity in this preliminary stage are virtues, not vices. However, the IEESA should avoid the fundamental economic errors built into some environmental accounting systems.

7. What Are Appropriate Techniques for Measuring Quantities and Values for Nonmarket Activities in the National Accounts?

One of the thorniest issues in developing augmented accounts involves measuring quantities and values for nonmarket activities. Chapters 3 and 4 of this report review techniques for measuring quantities and values in environmental accounts. The discussion in those chapters points out that estimates of the physical flows of these quantities are generally based on established scientific or business principles. For example, there are well-established principles for measuring and monitoring the volumes of petroleum and other subsoil minerals, the volume of timber, different soil types, exposure to pollutants, and concentrations of greenhouse gases. The difficulties with respect to resource and environmental quantities arise because there are generally no routine measures when these flows take place outside the marketplace. One of the key requirements of improved environmental accounting, therefore, is to improve these physical measures, particularly for environmental variables such as human exposures to pollutants. As is discussed in the next section, better measurement also would have important benefits for resource management and other public policies.

The largest conceptual issue that arises in extending the national accounts is how to value nonmarket activity. In the market sector, quantities are valued by their market prices, which reflect the valuation placed on marginal or "last" units purchased. Constructing nonmarket accounts that are fully consistent with market accounts requires finding proxies for marginal values in nonmarket behavior.

Environmental economists currently employ a wide variety of techniques in valuing nonmarket activities. Some rely on market activity or actual behavior. One example is the travel-cost method, which measures the value of a recreational site according to the time and other resources people expend to get there. A second behavioral approach, currently employed in the federal statistical system in both price indexes and the national output accounts, is hedonic analysis; under this approach, an activity is valued in terms of its components, such as when a computer is valued according to the implied market values of features such as memory and speed. Quite a different approach, relying on nonbehavioral data, is contingent valuation, which uses survey techniques to determine people's stated values for environmental or other variables, such as recreational sites or visibility at the Grand Canyon. Whatever the perceived strengths and weaknesses of these approaches, most specialists agree that non-market-value estimates have lower levels of precision, objectivity, and reliability than do hard market-based values, and much more validation of these nonmarket approaches remains to be done.

Techniques for valuation of nonmarket assets and activities are in their infancy, and new approaches and validation tests are now under way. As is true of new fields generally, there are fierce disputes, particularly about the validity and objectivity of nonbehaviorally based techniques such as contingent valuation. One major criticism of contingent valuation is that there is no budget constraint limiting the total expenditures on nonmarket activities to a total available amount. People's willingness to pay to save spotted owls or clean up Prince William Sound faces an unbounded psy-chic budget constraint on eleemosynary activities. Moreover, the task of embedding nonmarket valuation and contingent valuation in a larger double-entry bookkeeping system has received little research attention to date.

BEA takes a middle ground between a purist approach that uses only market prices and an aggressive approach that employs the best available estimates.[4] BEA holds that methodologies used to value nonmarketed goods and services must include constraints based on market and non-market inputs, including those involving time and income, and would use techniques that rely on reliable market and objective behavior. BEA may well rely on hedonic estimates of nonmarket values because these have been tested, because the agency has had experience with these approaches, and because they are based on actual market and nonmarket behavior. BEA is reluctant to rely on contingent valuation and nonbehavioral, will-ingness-to-pay approaches because they are not constrained to fit into a double-entry bookkeeping system and because their results are seen as implausible in many cases, inconsistent with the overall accounting frame-work, unstable when budget constraints are added, and extremely expen-sive to implement.

The panel is sympathetic with the reluctance of a government statistical agency responsible for producing the official national accounts to use con-

---

[4]The aggressive approach was used in a study of the benefits of clean-air regulations conducted by the U.S. Environmental Protection Agency (1997), which is reviewed in Chap-ter 4.

troversial procedures. Moreover, we recognize that nonbehavioral approaches such as contingent valuation have not been thoroughly calibrated and tested to ensure that they are reliable proxies for actual behavior. At the same time, the panel hopes further research will help resolve the uncertainties and provide sound and reliable methodologies for nonmarket goods and services. The payoff to developing comprehensive nonmarket accounts is great, yet without some method of valuing nonmarket activities and public goods, there will be major gaps in a comprehensive accounting system. Therefore, the panel recommends continued work in developing valuation tools that would be appropriate for a full set of augmented accounts.

5.7a The principles of physical measures of stocks and flows of many natural-resource and environmental assets and activities are reasonably well established. Generally, however, there are no routine measures when these flows take place outside the marketplace. One of the important requirements of improved environmental accounting is to improve such physical measures. These enhancements would yield substantial benefits in providing support for environmental and economic policies.

5.7b It has proven difficult to value many environmental and other nonmarket activities and assets. For natural-resource and environmental assets and activities, no single valuation method is free of problems or serves all the varied interests of potential users. Valuation methods used by BEA should rely on available market and behavioral data wherever and whenever possible. Although there are difficulties with nonbehavioral approaches such as contingent valuation, work on the development of such novel valuation techniques will be important for developing a comprehensive set of production and asset accounts.

Further research and validation on nonbehaviorally based techniques would be useful in order to determine their objectivity, stability, and reliability for national economic accounts (see recommendation 4.2).

8. What Should Be the Next Steps in Extending the IEESA?

A major decision involves the next steps in developing the environmental accounts. Before stopping work on the IEESA, BEA prepared a complete set of subsoil mineral accounts. It also undertook preliminary estimates of forest values, along with estimates for land underlying structures (see Chapter 4). In investigating other areas—recreational land, soil, wild fish, uncultivated forests, unproved subsoil assets, undeveloped land, air, and water—BEA found either data of questionable quality or no appropriate data on price or quantity.

Under BEA's phased work plan, assets such as forests that produce timber and vineyards that produce wine-grapes would be added. "Developed natural assets" such as oil, orchards, agricultural land, and forests would then be treated symmetrically with "made assets" such as houses, computers, and steel mills.

The panel agrees that improvements in valuing subsoil assets would be useful elements in a phased approach to environmental accounting. With respect to BEA's initial estimates for subsoil assets, the reported findings on the value of reserves—stocks, depletions, and additions—should be considered preliminary and tentative at this time. Improved accounts will require a better understanding of the value of mineral resources that are not now counted as known reserves, the impact of ore-reserve heterogeneity on valuation calculations, distortions introduced by the constraints imposed on mineral production by existing capital and other factors, and differences between the market and social value of subsoil mineral assets.

In the panel's view, the next priority under the phased approach should be sectors that include a significant aspect of market or near-market activity. Developing accounts for the commodity-producing value of forests is the obvious next step in developing the IEESA. Estimating the volume and value of forest timber appears to be relatively straightforward at this time, and the issues involved in the valuation are similar to those for subsoil assets. Another useful extension would be agricultural assets, particularly those involving livestock, vineyards, and land values and quantities. Beyond these sectors, the data become more problematic. Currently, data on fish stocks are unreliable because wild fish are fugitive assets, and there is no reliable census of the fishes. The panel did not investigate the water-resource sector in detail, but it determined that there are inadequate data on water stocks and water quality, and valuation of these resources remains a thorny issue because water value is highly variable depending on time, location, quality, and priority of right to usage.[5]

While recognizing the value of these phased incremental extensions, the panel reiterates that extending the accounts to include nonmarket activities is of the greatest substantive importance for augmented accounts. The panel's review indicates that accounting for environmental assets such as air quality is likely to have a major impact on consumption and investment. Developing environmental accounts is part of the even

[5]Water valuation issues are discussed in detail by the National Research Council (1997).

larger task of developing comprehensive nonmarket economic accounts. As noted earlier, the panel does not underestimate the awesome challenges involved in developing nonmarket accounts. Development of a set of accounts in this area involves major conceptual issues, the development of appropriate physical measures and valuation of flows and stocks, and constitutes a major scientific undertaking. As suggested above, the task of developing a comprehensive set of nonmarket accounts transcends the current scope and budget of BEA. Developing such accounts will require continued basic research on the underlying science and economics involved in estimating the benefits of public goods such as clean air, as well as applied research on accounting tools and valuation of nonmarket activities and assets.

5.8a If a phased approach is undertaken, the panel recommends that work to extend natural-resource and environmental accounting resume as soon as possible. Incremental improvements should focus primarily on developing those interactions between the economy and the environment that have market consequences. A useful step would be to refine estimates of subsoil mineral and timber accounts. Other incremental extensions should incorporate additional marketable assets and near-market goods and services—those that have close counterparts in marketed goods and services. There is a clear basis here for measuring quantities and establishing values for these market and near-market activities in a manner comparable to that used for the core accounts.

5.8b Construction of a set of forest accounts, focused initially on timber, is a natural extension for integrated economic-environmental accounts. The United States has much of the data needed for such accounts, and the analytical techniques are well researched. Other sectors that should be high on the priority list are those associated with agricultural assets, fisheries, and water resources.

5.8c While a phased approach to the development of environmental accounts is useful, a comprehensive set of natural-resource and environmental accounts will be critical to measuring the full impact of natural and environmental resources on long-term economic growth. Construction of a comprehensive set of economic accounts will require extensive research on the basic science and economics involved, as well as development of the appropriate tools for accounting and valuing nonmarket activities and assets.

BUDGETARY AND RESOURCE IMPLICATIONS

The cost to BEA and other agencies of constructing and maintaining the IEESA will depend on the intensity and extent of the effort. The costs would be small for a minimal program of small, incremental improvements limited to a few natural-resource sectors. Estimates from BEA indicate that the costs of such a small activity, including reinstatement of the pollution abatement survey, would be approximately $1.5 million annually.

It would be substantially more expensive to develop a full set of environmental and augmented accounts. In the long run such an effort would require developing a comprehensive accounting framework for exhaustible minerals and renewable resources along with a set of nonmarket service and investment accounts. Substantial incremental resources would be required both within BEA to develop the accounts and outside BEA to provide the data. Although the cost would depend on the extent to which BEA could draw on data and expertise from other agencies, it is likely that developing a full set of accounts would require incremental outlays for BEA and other agencies on the order of $10 million annually for a decade or more.

While noting the importance of developing a set of environmental and augmented accounts, the panel emphasizes that this work should not be done at the expense of the timely and current production of the current core accounts, along with improvements that reflect changes in the structure of the economy. As a result of several years of budgetary stringency, BEA has been hard pressed to maintain its current program, has been forced to curtail some of its activities, and has needed to be extremely selective in its choice of new initiatives. The agenda for improvements is extensive and includes many other important issues, such as improving the measurement of service outputs, improving measurement of international transactions, and accounting for stocks of and investments in human and knowledge capital. Maintaining the vitality of the national accounts while providing innovative and valuable new information is a worthy objective for BEA in the years ahead. Continued improvements in our data infrastructure are one of the soundest investments the nation can make.

DATA AND RESEARCH NEEDS FOR IMPLEMENTING ENVIRONMENTAL ACCOUNTS

In its charge, the panel was asked to "compare methodologies with research in other countries and in non-governmental research . . . and recommend improvements and research needs." Extending the NIPA to include the economic impacts of resource and environmental flows and assets would require considerable upgrading of the national database in these areas. This section addresses issues related to data collection and design.

Need for Interagency Cooperation on Data Collection

As noted in Chapters 3 and 4, much valuable information necessary for integrated environmental and economic accounts is already collected by the federal government and is potentially available to BEA. Extensive information is available in federal agencies on physical stocks and values of economically important natural resources, including subsoil minerals, energy, timber, commercial fisheries, and land. BEA's preliminary work on the Phase I accounts made use of existing data on the physical quantities and market values of such natural-resource assets. However, much of the data necessary for developing environmental accounts is currently unavailable or insufficient. One important step, therefore, would be to undertake a focused effort to increase and improve the data necessary for this work. Without significant improvement in this area, development of a full set of empirically based environmental accounts would be impossible.

Fortunately, much of the information needed to construct and maintain environmental accounts would also be useful to other federal agencies with resource management responsibilities. This is particularly the case for natural assets under federal stewardship. For example, better information on the value of minerals on federal lands and the net value of minerals extracted from federal lands would be useful in determining appropriate royalty rates and patenting policies for resources not allocated through competitive auctions. The same information would be useful to BEA in constructing environmental accounts for exhaustible natural resources.

In the case of renewable resources, better information on the stumpage value of timber in national forests would be useful not only for accounting purposes, but also for better management of these forests and for the difficult decisions required on the balance of different uses, including timber harvesting, wilderness preservation, watershed management, and recreation. Better information on fish stocks, depletion of fish stocks, and resource values net of extraction costs would be valuable to the National Marine Fisheries Service and to the Fisheries Management Councils and would also support U.S. negotiations in international fishing treaties. These agencies have been hamstrung in their efforts to prevent overfishing by a lack of reliable information on changes in stocks of commercial fisheries and on the dissipation of fisheries rents.

In the case of environmental resources such as air and water quality, better information on the economic value of marginal changes in air and water quality, which would be essential for constructing a comprehensive set of environmental accounts, would also be essential for sound benefit-cost analyses that the U.S. Environmental Protection Agency (EPA) is required to undertake in regulatory decision making. One of the most serious weaknesses in the U.S. environmental database is the lack of comprehensive and reliable data on actual human exposures to major pollutants. Better information on physical emissions trends, human exposures, and the economic impacts and damages due to air and water pollution would be valuable for expanded accounting measures of productivity. In summary, there are strong synergies between BEA's data needs for implementing its environmental accounts and other agencies' data needs for resource and environmental management.

Consequently, there would be great value in a cooperative and coordinated approach among federal agencies to the collection and management of improved natural-resource and environmental data. Definitions and coverage of existing surveys could be modified at relatively small cost to meet the needs of the environmental accounts while also providing better data for policy making. Raw data could be formatted and processed in more than one way to serve multiple purposes. Useful data collection efforts that might be found expendable by one agency operating under tight budgetary constraints might be continued under cost-sharing agreements among several agencies. Existing statistical coordinating and advisory bodies within the federal government, including the Office of Management and Budget, could play a useful role in coordinating data collection efforts useful for both environmental accounting and other important federal purposes.

In addition to coordination of data collection and management efforts, there is also a need to coordinate standards for accounting and measurement. Even though the general conceptual basis for environmental accounting is reasonably well established in theory, many issues arise in constructing the empirical counterparts to general concepts. Estimation methods that are equivalent in theory will typically yield different empirical results when used with actual data, and choices must be made among alternative valuation methods and data sources. Work on the valuation of natural resources under federal control is ongoing under the auspices of the Federal Accounting Standards Advisory Board. Close cooperation among BEA, other federal statistical agencies, and private researchers would be important for providing estimates of quantities and values that are appropriate for national-income accounting.

5.9 Extending the national accounts to include a full set of natural-resource and environmental impacts would require a major, focused

effort to improve the databases on quantities and values of key natural resources and environmental variables. Without significant improvement, it would not be possible to develop a full and reliable set of empirically based environmental accounts. Much of the information needed to construct and maintain environmental accounts would be highly useful to other federal agencies, particularly for natural assets under federal stewardship and for environmental activities for which the federal government has responsibility to undertake benefit-cost analysis. A cooperative and coordinated approach among analytic teams of researchers from different federal agencies and the private sector to collect, analyze, and manage improved natural-resource and environmental data would be valuable not only for developing natural-resource and environmental accounts, but also for promoting better monitoring, assessment, and policy making in these areas.

Data and Research Needs with Respect to Exhaustible Resources

BEA's preliminary implementation of its environmental accounts resulted in estimates of accounts for subsoil minerals, including fuels, metals, and nonmetallic minerals. In its 1994 article on minerals accounting (1994b), BEA addressed a number of data and implementation issues. Information on production, production costs, reserves, and reserve changes is less complete and accessible for most nonfuel minerals than for fossil fuels. Standardization of classifications among data collection agencies could improve the information base.

All the valuation methods attempted by BEA in Phase I—reviewed in Chapter 3 of this report—are approximations to ideal measures of the market value of subsoil resource stocks and flows. These approximations are required because the information needed to implement ideal measures is unavailable. Implementing ideal measures of resource values based on the discounted present value of returns generated over the life of the resource would require projections of future prices, quantities, and discount rates. However, better approximations could be obtained with additional research and information. The most important topics include the following.

The heterogeneity of resources. Resources actually utilized, for which market data are available, tend to have the highest quality and lowest cost of those currently available. The unit value of additions to reserves may differ substantially from the unit value of extracted or harvested reserves. This is true both for exhaustible resources and for renewable resources, such as timber. Valuing additions to reserves or the entire body of reserves at the same price as resources currently extracted or harvested may seriously bias estimates of the value of the stock.

Information is potentially available on the distribution of many deposits of ores and mineral fuels by grade, depth, and other relevant characteristics. Similarly, information is available about the characteristics of standing timber stock by species, grade, accessibility, and age. Fish biologists have information about the size of the recruitment class added to a fish population in a given year. Such information could be used to refine the estimates of stock values and of changes in the stock over time, and could provide substantially more accurate estimates of the value of additions and depletions to the stock of resource assets.

Unproved reserves and resources. Under current approaches, only the value of proven reserves is usually included in the product and asset accounts. Proven reserves are, by definition, those resources which are known with reasonable certainty to be economical to produce at current prices and using currently available technology. Because unproven or speculative resources may be produced in the future as prices rise and technologies improve or as potential reserves are developed, they may have a market value. Although BEA has indicated plans to produce such estimates in the future, they are not included in current accounts. Further information on the value of unproven resources could be obtained from such sources as bids on offshore oil and gas leases.

Some mineral and timber resources, though known, are not commercially available because they occur on federal or state lands that have protected status. These resources also have an option value because their legal status may change. For example, the federal government recently sold the Elk Hill petroleum reserve. Information on the extent of such resources, if made available for production purposes, could be obtained from federal land and resource management agencies.

Value of associated capital. Mineral reserves usually consist of mineral assets and associated physical capital constructed to exploit the reserves. It is necessary to estimate the value of the associated tangible capital in order to estimate the value of the natural-resource stock or flow (see Chapter 3), Otherwise, the estimated resource values may be substantially overstated. Though BEA has attempted to make such provisions, further information gathering is needed to refine these estimates. For example, Chapter 3 examines techniques for improving the simplest Hotelling valuation approach by incorporating a measure of the value of the physical capital constraint on production. Consequently, more empirical information is needed on the extent to which production of oil, gas, and nonfuel minerals is likely to be limited over short time periods by physical capital constraints. Such research would allow a better estimate of the value of associated capital.

Liabilities associated with resource extraction. For institutional reasons, mining historically has not provided private firms with adequate incentives to forestall or remedy many environmental effects. Consequently, there are thousands of active and inactive mine sites responsible for environmental harm to surrounding properties through leaching, subsidence, or visual impairment. Such sites could be termed natural-resource liabilities. Currently, there is no proper accounting either for the stock of such liabilities or for the change in their value. Data are available from federal oversight and regulatory agencies to account for such liabilities, and may also be obtainable from mineral leases that specify restoration once mining operations have been completed.

Regional disaggregation of resource accounts. BEA's preliminary estimates indicated that the value of exhaustible resource stock changes does not constitute a large fraction of national net capital formation. Nonetheless, such changes do represent substantially larger fractions of net investment in particular regions or individual states whose economies are relatively resource-dependent. For example, extractive and other resource-based industries are economically significant in Alaska, the mountain states, the Northwest, and parts of the South and Northeast. Within a framework of supplemental accounts, it would be useful to present regionally disaggregated resource accounts. Doing so could create additional data requirements. Since the underlying source data on production and stocks are generally collected for states and counties, the main requirement is that the locational tag not be lost in the process of data aggregation.

In improving BEA's accounts for subsoil assets, further analysis is needed to assess different valuation techniques. Preliminary assessments indicate that the standard Hotelling valuation approach overestimates resource values, and this finding should be incorporated in valuation approaches. Further work is necessary to determine the importance of heterogeneity of reserves, the value of unproven and speculative assets, the value of associated capital, the liabilities associated with resource extraction, and the regional impacts of activities associated with subsoil assets. Where the costs are reasonable, BEA should develop and report regional data on important natural-resource and environmental activities, such as those for subsoil assets. The recommendations of the panel in this area are contained in Chapter 3. See particularly recommendations 3.2 through 3.7.

Data and Research Needs for Accounting for Renewable Resources

Asset values. BEA's plans for developing the environmental accounts include making estimates of developed natural assets such as timber in managed forests, cattle, vineyards, orchards, cultivated fish stocks, and developed land. In a later stage, BEA would account for uncultivated biological resources such as wild fish, timber and other plants in unmanaged forests, and other uncultivated biological resources. The construction of accounts for agricultural, horticultural, and animal husbandry assets poses no major data issues, and the U.S. Department of Agriculture, together with the U.S. Bureau of the Census, has a comparatively full set of information on these issues. Similarly, data sources, though of varying accuracy, are available from which to estimate the market value of developed land.

Accounting for renewable resources such as forests encounters some of the same information issues and data gaps as does accounting for exhaustible resources. Managed forests other than plantations contain trees of heterogeneous ages, species, and other characteristics. Harvested trees will generally differ in unit value from the unharvested stock and from additions to that stock. Data on the heterogeneity of timber stocks are particularly important because harvesting is likely to be limited to the more valuable stocks, and therefore stumpage price estimates derived from such commercial operations cannot be reliably extrapolated to other unexploited stocks.

Though the national forests contribute a small share of total harvested timber, there are particular problems in accounting for wood extracted from these forests. Though standing timber is typically sold through auction bids, sales prices will not represent the market stumpage value of the timber for those sales that have only a single bidder. In such sales, the winning bid usually corresponds to the Forest Services's administratively determined minimum acceptable bid. Bids are also influenced by cost considerations. Logging contractors are required to operate under conditions imposed to protect other multiple-use environmental values, such as water quality, habitat protection, and recreational and aesthetic values. These conditions may increase logging costs and therefore reduce the amounts potential contractors are willing to bid for logging rights. Offsetting these upward pressures on costs in the national forests, the government bears some logging costs, notably those of road construction, which are financed out of road credits. Research will be necessary to determine whether transaction data based on bids for logging rights in national forests are an accurate source of information on stumpage values, or whether they would require some adjustment to be useful in the environmental accounts.

With respect to timber harvested on private lands, difficulties arise in allocating joint production costs in industrial forestry operations carried out by integrated pulp and paper or forest product companies. A substantial fraction of total timber harvested originates on lands owned and operated by such companies. In addition to problems of joint cost allocation, there are also problems of establishing or inferring prices for logs that are not bought or sold but processed by integrated companies into final products. Further issues arise with respect to valuation of timberland, as opposed to the standing stock of trees. In its initial effort, BEA assumed that timberland, on average, is worth as much as agricultural land. BEA reasoned that if not worth at least that much, timberland would be converted to agriculture, which may be its next-best use. However, the opposite might also hold true—that timberland is kept in forest because the land is not worth converting to agriculture. Better region-specific information on the capabilities and market value of forested land would be helpful in improving the estimates.

Measurement of service flows. The main challenge for research and data collection arises from the need in a comprehensive set of environmental accounts to estimate the environmental service flows provided by forests, freshwater, and other renewable resources. Because use patterns have historically been dominated by commodity production for the marketplace (such as agricultural production using land and timber production from forests), there is much more data available on commodity production values than on environmental service values. Nonetheless, economic research indicates that many renewable resources, especially those in the public domain, are today more valuable as sources of environmental service flows than as sources of marketed commodities. Ignoring service values would therefore substantially distort asset and production accounts.

There are many useful data sets on the use of publicly held renewable resources for recreational purposes. For example, the government collects data on the number of visitor-days for recreational purposes to national forests, public beaches, and other protected areas. Economic research has estimated service values and related those values to various qualitative aspects of the services, such as congestion, water and air quality, and visual characteristics. This research is based on methodologies developed by environmental economists. Some such methodologies derive estimates of values from observations of market or behavioral decisions, such as travel costs incurred to participate in recreational activities. Such information can be used to estimate the value of current service flows provided by renewable resources and the contribution of these service flows to the underlying asset values.

Problems can arise in the use of current estimates. Care must be taken to ensure that the values are marginal or incremental values, rather than total or consumer-surplus values. Many studies include consumer surplus and are therefore not comparable to the price and value approach used in the current national accounts. Moreover, the establishment of either values or quantitative estimates of environmental service flows related to such ecological functions as wildlife habitat, nutrient recycling, carbon sinks or sequestering, biodiversity, and hydrological regulation is still highly speculative. Inclusion of such estimates in the national accounts is questionable today and might be postponed until data and methodologies in this area are improved.

More research is needed on the effect of stock changes on the value of these service flows because the relationship is complex and current information may be inaccurate. For example, a reduction in standing volume of timber may change water outflows from a forest, increase habitat for some animals and decrease habitat for others, and increase some kinds of recreational services while decreasing others. Storage and diversion of waterways for irrigation purposes may likewise provide habitat for some aquatic species and destroy it for others, and increase some recreational uses but eliminate others.

Many of the same issues arise in accounting for the market-related functions of renewable resources and subsoil assets. Much work already exists on valuation of forests and timber, but further research on valuation is necessary to determine the accuracy of the Hotelling approach. The major challenge in estimating both asset values and service flows lies in determining appropriate values for nonmarket aspects, which are particularly important for forests. Recommendations for forests are in Chapter 4 (see particularly recommendations 4.5, 4.8, and 4.9).

Accounting for Changes in Air and Water Quality

Developing improved accounts for environmental assets such as air and water quality or nonmarket services of natural-resource and environmental assets is an important goal of augmented accounting. Accomplishing this goal involves both measurement of the costs of pollution abatement and estimates of the value of the market and nonmarket services provided by these assets. One important initial step undertaken by BEA was the construction of a set of estimates of pollution abatement and control activities. This effort has unfortunately been discontinued because of budget cuts imposed on BEA. These estimates are an important aspect of any economic assessment of the environment.

The development of accounts for changes in air and water quantity was postponed to Phase III of the IEESA effort, as was accounting for uncultivated biological resources such as wild fish and undeveloped land. Though ambient environmental quality represents an important dimension of current consumption and from a conceptual point of view belongs within an expanded set of environmental accounts, data needed to implement this approach are currently unavailable except in a small number of cases.

Data on air and water pollution illustrate the difficulties. Although

EPA often conducts benefit-cost analyses to support regulatory decision making, the resulting estimates of the economic value of marginal changes in environmental quality are typically limited to a limited class of pollutants, pollution sources, and geographical areas. They cannot be readily extended to the more comprehensive national estimates needed for a set of augmented accounts. Moreover, they usually examine the incremental costs and benefits of a regulation and seldom calculate the total damages or changes in damages from a historical or normative baseline. Finally, for the most part, the valuations of benefits contained in these studies are not derived from market transactions or behaviorally derived values. Unless EPA and other agencies undertake or underwrite a substantial effort to improve the data in this area, the lack of comprehensive information on the value of nonmarketed environmental goods and services is likely to constrain the development of a full set of environmental accounts.

The nub of the difficulty in constructing a set of environmentally adjusted national accounts lies in estimating the consumption services of environmental assets. Deriving such estimates through the conceptually correct "damages borne" approach—which measures the actual damages or impacts of changes in environmental flows—would require improved data on ambient air and water quality, which vary temporally and spatially, and on the profile of exposures of humans and other organisms to pollution. Perhaps the most important lacuna is data on actual human exposures to air pollution, which are virtually absent from the U.S. national data system.

Economic damage assessments—whether based on epidemiologically estimated dose-response relationships or more directly on hedonic property, wage, or travel-cost studies—do not now constitute an adequate empirical base on which to construct environmental accounts. Estimates are sensitive to specification and data and tend to be time- and location-specific. Moreover, econometric estimates based on compensating and equivalent variations often give substantially different results. Additional research on the valuation of pollution damages is needed, with special emphasis on the value of marginal changes in environmental quality. Research on so-called "benefits transfer" techniques, which allow damage estimates to be adapted to other populations and pollution concentrations, is also needed. For these reasons, implementing Phase III of BEA's proposal would require a considerable research component.

Finally, two recommendations presented in Chapter 4 are worth reiterating here. First, BEA's annual survey of pollution control and abatement expenditures should be reestablished (see recommendation 4.7). Second, the nation needs improved measures of physical indicators for many environmental variables, particularly those involving human exposures. In the designing of environmental indicators, policy issues should dictate the choice of variables and the focus of the research. Measures should focus on human health and on social, economic, and ecosystem effects, rather than simply on pollutant concentrations or similar intermediate variables (see recommendation 4.3).

Frequency

The panel considered the issue of the frequency of estimation and publication of natural-resource and environmental accounts. Because the underlying physical activities generally take place at a slow pace, particularly relative to business cycles, it is not sensible to aim for reporting in the satellite accounts more frequently than on an annual basis. Annual reporting is recommended for those activities—particularly subsoil assets and forests—for which annual data are readily available. For other measures, including input-output analysis, measures of comprehensive or sustainable income, and similarly complex constructions, quinquennial reports may be a reasonable goal. Frequent analysis and reporting are not necessary given the source data, costs, and temporal evolution of assets and activities that are being measured. Neither the data nor the likely uses of such accounts would suggest the need for monthly or quarterly data, particularly given the problems created by the short-run volatility of mineral commodity prices.

5.10 The panel recommends regular periodic accounting in the natural-resource, environmental, and other augmented accounts. The accounts for subsoil assets and forests could be developed, calculated, and reported on an annual basis. For other measures, less frequent accounts, perhaps quinquennial, would be appropriate and cost-effective.

APPENDICES

APPENDIX

A

Sustainability and Economic Accounting

 In light of increasing environmental problems in many sectors, concerns have been raised about the sustainability of current patterns of economic activity in both developed and developing countries. What are the environmental, social, and economic implications of continuing "business as usual"? Will the current path of population, energy use, and growth of human settlements do irreversible harm to the natural ecosystems and life-support systems of the globe? Are we headed for economic overshoot and collapse if we continue to rely on today's technologies? In short, is our economy on a sustainable path?

The concept of sustainability was popularized by the report of the Brundtland Commission, which defined "sustainable development" as "development that meets the needs of the present without compromising the ability of future generations to meet their own needs" (World Commission on Environment and Development, 1987:43). The concern addressed by the Brundtland Commission was whether nations are wasting or abusing their natural endowments of minerals, soils, forests, and aquifers, along with vital environmental resources such as clean air and water, as well as the stock of genetic material.

A parallel effort among economists has been the development of measures of national income and output that take notions of sustainability into account. This appendix examines issues of sustainability from an economic point of view and describes the relationship between measures

183

of sustainable income and augmented national income accounting. It reviews alternative definitions of income and output; shows how net national product (NNP) is a measure of sustainable consumption under idealized conditions; and then demonstrates how the linkage between current output measures and measures of sustainable consumption breaks down in the presence of nonmarket consumption and investment in environmental, human, and technological capital.

CURRENT PRODUCTION VERSUS SUSTAINABLE CONSUMPTION

The origins of the concept of "social income" or "national income" go back centuries, but two fundamental approaches can be distinguished— one based on the idea of current production and one based on sustainable consumption. The former is the basis for modern national income accounting, while the latter is often used as the appropriate concept for the measurement of sustainable income.

Production-based measures. Those who constructed the earliest national accounts were concerned with obtaining accurate measures of national output and national income. Particularly important was tracking current production so governments could take measures to stabilize the business cycle. In attempting to develop a careful definition of national income and output for production-based measures, economists have usually relied on the concept of *Hicksian income*, which defines net national output as the maximum amount that can be consumed while leaving capital intact (see also Chapter 2).[1] In practice, this means national output is defined as consumption plus net investment. This concept is the standard definition of NNP used in the national income accounts of virtually all nations today. It is production based in the sense that it measures production in a given period in terms of market prices. Such a measure is not concerned with whether the economy is sustainable or not, whether production and consumption are growing or declining, or whether the economy is on a path toward prosperity or extinction. Rather, it measures what consumption would be if net investment were zero (that is, if the capital stock were kept intact), measured at the market prices of the economy. Given this definition of income and output, it is easy to understand the rationale of current approaches to augmented accounting. The purpose of these extensions is to expand the purview of the accounts to

---

[1]The basic reference is Hicks (1939:173, 178), who defined his production-based measure as "the maximum amount which can be spent during a period if there is to be an expectation of maintaining intact the capital value of prospective returns . . . ; it equals Consumption plus Capital accumulation"; see also Hicks (1940) and Kuznets (1948a, 1948b).

include a broader definition of "capital." These studies augment conventional national income by including corrections for human capital; government capital; the stock of research and development; and natural capital such as forests, mineral resources, and environmental resources.

Sustainability-based measures. While standard production-based measures of income are useful tools for measuring current production, they do not address concerns about the sustainability of current decisions. An alternative approach, emphasizing sustainability, is provided by Solow (1992), who suggested in his talk of 1992 that a sustainable path for the national economy is one that allows every future generation the option of being as well off as its predecessors. Similarly, according to Repetto (1986:15-16), "The core of the idea of sustainability, then, is the concept that current decisions should not impair the prospects for maintaining or improving future living standards." For purposes of the present discussion, *sustainable national income* is defined as the maximum amount that can be consumed while ensuring that all future generations can have living standards that are at least as high as that of the current generation.

It should be emphasized that the sustainability-based approach—while deemed particularly useful and appropriate in the context of designing comprehensive national income accounts—is but one of many approaches to analyzing the sustainability of an economy or of the interactions between the economy and the environment. Literally dozens of definitions and approaches have been used in different contexts. It will be useful for present purposes to discuss one major distinction among the different approaches, which relates to the degree of specificity of the variables or objectives to be sustained.

The economic approach to sustainability considers only economic activities and excludes many important individual and collective activities and processes. Economic welfare in this view consists of per capita consumption of goods and services, both market and nonmarket. Living standards are measured on a per capita basis. Consumption includes market items such as food, shelter, and entertainment; in principle, it also includes nonmarket items such as home-cooked meals, along with recreational activities such as fishing or gardening. Consumption does not include many other important values, however. It excludes political and social values such as the importance of fairness, of freedom of speech or association, of religious convictions, and of happy families. Moreover, the values considered are ones that originate in human values. Thus, while human concerns and values about the natural environment are included, the feelings of animals or any intrinsic value of natural ecosystems, such as the existence of coral reefs or of species, are not. To exclude these latter measures is not to deny that they may have value; rather, our economic measures cannot go beyond the boundary of measurable economic activities.

Moreover, economic analyses of sustainability examine consumption or sustainability at the highest level of aggregation—the level of average consumption today and in the future. This level of aggregation masks a number of important ways of disaggregating the complex ensemble of economic and environmental activities. It omits details such as the sectoral or asset breakdown (for example, the separate trajectories of reproducible capital, stocks of subsoil minerals and timber, the quality of air and water, the health of salt marshes, and the value of stocks of genetic material). It assumes that within a sector or asset class, substitutes (including technology) will replace high-priced goods and services. It also overlooks the distribution of consumption among different groups within a country or among countries. It does not distinguish among different future generations and focuses only on the present generation versus the generalized future.

In addition, most treatments of sustainability do not deal with issues of uncertainty or risk. Must a path be sustainable with absolute confidence, or on average, or 90 percent of the time? Would we prefer a certainty of nondeclining consumption over an alternative that involves a robust growth in living standards plus a tiny chance of a small decline? How would we feel about a promising technology that offers a 99.9 percent probability of sustainability and a 0.1 percent chance of extinction? A short journey down the road of stochasticity raises numerous unanswered questions about the concept of sustainability.

In limiting the present analysis to this highly generalized and aggregated version of sustainability, it is recognized that many worthwhile goals will be overlooked. An alternative view of sustainability, for example, might hold that "maintaining capital intact" should apply at a more disaggregated level than the entire asset base of an economy. This narrower perspective might hope to protect certain assets or flows or subsystems—such as a suite of species or a group of important ecosystems, or even "natural capital" more generally—so that future generations could enjoy them at today's levels. Such a perspective depends on "sector-specific" and "use-specific" definitions of sustainability. Defining sustainability in this narrower sense is often useful as a guide to policy making or as a practical shorthand way of expressing certain desirable conservation goals, but it generally is too narrow and subjective to serve as a principle for constructing measures of national income. Moreover, if taken literally, the injunction to keep "natural capital intact" is probably infeasible because human activities inevitably cause the levels of some natural assets somewhere to decline.

Sector-specific or use-specific definitions of sustainability raise other practical problems. One issue is selection of the list of assets to be maintained. Which specific set of resources is to be maintained? Who selects this list? Who decides on global assets? Answers to these questions matter a great deal because, both literally and figuratively, what is sustainable for the forest is unsustainable for individual trees. Additionally, from a technical point of view, the sector-specific or use-specific approach assumes that no substitution is allowed between the particular resource or use chosen to be sustained and other resources not on the selected list. If killer whales are on the sustainability list while the porpoises they eat are not, not one more killer whale can be harvested, even if doing so would allow one thousand more porpoises to live.

This short discussion should help indicate both the usefulness and the pitfalls of alternative definitions of sustainability. On the whole, despite its shortcomings, the broad measure of sustainability is likely to be the best single criterion for defining sustainable national income. This measure defines sustainable income as the maximum amount that a nation can consume while ensuring that all future generations can have living standards that are at least as high as that of the current generation. Such a broad concept provides an intuitively appealing way of providing a simple summary answer to the complex question of where our economic growth and development are taking us.

CORRESPONDENCE BETWEEN NET NATIONAL PRODUCT AND SUSTAINABLE INCOME UNDER IDEALIZED CONDITIONS

What is the relationship between concepts employed in the current National Income and Product Accounts (NIPA) and measures of sustainable income? To begin with, it should be noted that the most popular measure of output, gross domestic product (GDP), differs from a conceptually appropriate measure, NNP, in two ways. First, GDP includes capital consumption (see glossary), which leads to double counting of this source of income. Traditionally, output measures have emphasized gross rather than net product because depreciation is difficult to measure accurately. Second, domestic product excludes the net factor earnings abroad of domestic residents, which is included in national product. Inclusion of net factor earnings abroad is desirable if output is designed to measure the sustainable consumption of the nation. The recent switch in emphasis from national to domestic product came about because domestic product is more closely related to domestic output and employment. While the emphasis on GDP rather than NNP is understandable, the panel emphasizes that NNP is conceptually preferable as a measure of sustainable income.

However, there is a close relationship between traditional measures of output and ideal measures. This relationship, known as the *output-sustainability correspondence principle*, holds that under idealized conditions, NNP and sustainable income are identical. More precisely, when the national accounts include all stocks of capital and other dynamic features that affect production and when markets accurately capture the social values of all inputs, NNP is an appropriate measure of sustainable income. In other words, the sum of total consumption and net capital formation is equivalent to the maximum sustainable amount of consumption an economy can indefinitely maintain. Hence under idealized conditions, including zero population growth, extending the NIPA toward a comprehensive measure of Hicksian income would make output and income more accurate indexes of sustainable income.[2]

The balance of this section is devoted to explaining the output-sustainability correspondence principle; a number of qualifications to the principle are presented in the next section. We simplify the analysis by assuming that there is just one composite consumption index, which measures the real standard of living of the representative household. This generalized consumption index should be interpreted as being broader than a traditional index of real consumption because it comprises both market and nonmarket consumption, including not only food and concerts, but also wilderness experiences and highway congestion.

Consumption is produced by a large number of different kinds of capital goods. Some of these goods, such as equipment and structures, are included in the national accounts. Others are nonmarket capital, such as stocks of mineral deposits and fish, human capital, technological capital in the form of patents, and the like. Ideally, the list of capital goods should be as comprehensive as possible, subject to the limitation that the goods have meaningful units of measure and that scarcity prices—either actual market prices or imputed shadow prices—can be calculated to measure their values.

The major behavioral assumption here is that the outputs and prices of the economy are generated by an optimized economy for which there is a complete set of accounts and in which all spillovers are internalized. This means that consumption, investments, and prices are generated by a process in which (1) the accounting system is complete in the sense that all dynamic elements and capital stocks—natural, environmental, and technological—are included in the measure of income; (2) all transactions are "internalized," meaning that markets capture all the social costs and benefits of all economic activities; and (3) output, consumption, and investment result from social decisions that optimize a consistent intertemporal objective function.

---

[2]This proposition dates back to Weitzman (1976). For a recent comprehensive treatment of the subject, see Aronsson et al. (1997). For a dynamic view see Perrings (1998).

The mathematical proof of the correspondence principle depends on a multisector generalization of the Ramsey-Cass-Koopmans optimal growth problem. In this approach, the objective is to maximize the present discounted value of the utility of consumption. The fundamental relationships are a utility function that represents society's intertemporal preferences over alternative consumption streams, a production function that indicates how consumption can be produced as a function of a wide array of capital stocks and autonomous dynamic factors, and a pure rate of time preference that indicates the relative priority of consumption of different periods or generations. For this purpose, it is not necessary to observe the utility function or the rate of pure time preference. Rather, it is assumed that the economy behaves *as if* it were the solution of an optimal growth problem. The present discounted utility specification has been axiomatically derived by Debreu and Koopmans as the appropriate intertemporal welfare function from postulates that encompass a general set of social objectives (for a full discussion of this approach, see the references in footnote 2).

Under these conditions, along with no growth in population, comprehensive net domestic product (NDP) is an appropriate measure of sustainable income. That is, national income as measured by current comprehensive consumption plus the sum of the values of the net accumulation of assets in different sectors is equivalent to the maximum level of consumption that can be indefinitely sustained. Moreover, under the stringent conditions of a complete, internalized, optimal growth path, this measure of sustainable income will be exactly captured in measured NNP. One important conclusion is that to the extent that the national accounts omit important components of consumption and of net capital accumulation, they may provide misleading measures of sustainable income.

QUALIFICATIONS TO THE OUTPUT-SUSTAINABILITY CORRESPONDENCE PRINCIPLE

The output-sustainability correspondence is of fundamental importance for guiding decisions about the design of the NIPA. However, there are important practical and theoretical qualifications to this principle that must be emphasized. These qualifications concern (1) the incompleteness of the consumption and asset categories; (2) the presence of autonomous technological and other processes; (3) revaluation issues; and (4) problems associated with imperfect markets, imperfect foresight, and other departures from the optimal-growth framework. Any of these four conditions generally implies that measured NNP will depart from the ideal measure of sustainable consumption.

Incompleteness

In measuring comprehensive national income and output, we desire that our accounting system be as complete as possible in the sense of including as many components of consumption and net investment as is practical. To the extent that we omit certain items, this will lead to errors or residuals in our measure of national income. To clarify this point, we can write augmented NNP as:

$$\text{augNDP}_t = C_t^I + \Delta K_t^I + C_t^{II} + \Delta K_t^{II} = \text{sustainable income}_t, \quad (A.1)$$

where augNDP_t is augmented NNP and is equal to consumption and net investment, C_t^I and C_t^{II} are consumption of types I and II, and ΔK_t^I and ΔK_t^{II} are net capital formation of types I and II. Suppose that type I consumption and capital formation refer to those flows as measured in the standard NIPA, while type II refers to nonmarket activities not captured in the accounts, such as the flows associated with forests, fisheries, and underground aquifers. Moreover, suppose these two sectors are all that matter for economic welfare. If the assumptions of the correspondence principle hold, one can accurately measure sustainable income by adding the appropriate values for sector II to conventional NDP. If, by contrast, sector II is omitted, sustainable income will differ from national output by a residual term as follows:

$$\text{sustainable income}_t = \text{NDP}_t + C_t^{II} + \Delta K_t^{II} = \text{NDP}_t + R_t^A \quad (A.2)$$

In equation (A.2), sustainable income can be measured as conventional NDP plus a residual (R_t^A), which is equal to omitted consumption and investment. The residual R_t^A in equation (A.2) reflects the omitted consumption and investment terms, $R_t^A = C_t^{II} + \Delta K_t^{II}$. The residual element in R_t^A reflects elements of economic welfare that are not reflected in measured national output. Naturally, this residual may be negative or positive, depending on whether the sum of nonmarket consumption and net investment is negative or positive.

The major point here is that because of incompleteness in the consumption and investment categories, conventional measures of NDP will depart from sustainable income. Moreover, given the vast array of nonmarket activities—from leisure and home-based investments in human capital to nonmarket consumption and investments in environmentally sensitive sectors—there is a strong presumption that the residual may be significant and that current measures of national output do not adequately reflect sustainable income.

Technology, Institutions, and Other Autonomous Processes

In addition to market and nonmarket capital stocks, there are likely to be important social and technological processes that affect the trend of production. This point is well established in studies of economic growth and development, which have found that conventional capital formation explains only a small fraction of the growth of individual nations or the differences among nations. Examples of other important factors are the level of and improvement in technology; the institutional and legal structure, including the strength of tangible and intellectual property rights; the stability of the political system; the level of openness of the economy and the presence of pacific or warlike neighbors; and the honesty or corruptness of public and private transactions. It is likely that many environmental factors—such as the value of the biosphere or the climate—fall into this category.

These factors are in some sense "social capital" and clearly affect a nation's productivity and future income. But to call them "capital assets" is really a metaphor; they cannot be treated as such in any serious accounting sense. There is no metric for measuring many of these important social elements, nor is there an established methodology for valuing them. From the point of view of the national accounts, they are autonomous dynamic factors that may have a substantial impact on productivity.

These autonomous factors also lead to a divergence between sustainable income and measured national output. For example, if ongoing technological change leads to sustained productivity improvement, future generations will have higher levels of income than current generations, and current sustainable income is therefore higher. If, by contrast, current activities are leading to a general deterioration in the absorptive capacities of the environment or if climate change will lead to irreversible damage to the fertility of the earth, current sustainable income is lower.

Relatively little work has been done on the size or sign of the missing residuals, particularly those due to autonomous dynamic factors. Work on the sources of economic growth indicates that technological change has historically been a dominant factor in the growth of measured per capita output and living standards. Recent studies also point to the importance of institutional arrangements, such as the openness of the economy, the role of markets in resource allocation, and proximity to coastlines or large and growing markets. These findings, along with illustrative calculations, suggest that the autonomous residual may be positive and relatively large. Much work is needed in this area to obtain a better appraisal of the importance of these autonomous dynamic factors.

Revaluation Effects

Revaluation effects, or real capital gains and losses, raise perplexing problems for measuring income and output (see also Chapter 3). Conventional measures of real national output involve only the level and changes in the weighted quantities of goods and services consumed and invested; revaluation effects are omitted from conventional income and product measures. A difficulty arises when there are changes in the relative prices of consumption goods across borders or over time. The simplest example occurs in the case of a change in a country's terms of trade. Suppose there is a 10 percent permanent and unexpected rise in world oil prices relative to consumption goods. An economy that produces only oil will have a 10 percent increase in its sustainable consumption level even though there has been no change in the time path of its physical oil production. Similar effects would occur if there were changes in real interest rates, which are in effect changes in the terms of trade between the present and future.

Changes in prices lead to revaluation effects, which are changes in the value of income or capital in terms of current consumption goods. Inclusion of revaluation effects in sustainable income is a major departure from traditional definitions of output in the national income accounts. In fact, from the point of view of a small open economy, price changes are often as important a determinant of consumption possibilities as changes in investment, in trade regimes, or in technology. In principle, current approaches to measuring sustainable income treat revaluation in a way that is parallel to the treatment of autonomous dynamic influences. (In practice, such revaluations are rarely done in national income accounting.)

Revaluation effects and other dynamic factors add another term to equation (A.2) that reflects the positive or negative contribution of price and interest-rate changes and other dynamic effects to the highest level of sustainable consumption. Equation (A.2) can be modified to incorporate revaluation effects and the influence of the autonomous elements as follows:[3]

$$\text{sustainable income}_t = NDP_t + R_t^A + R_t^B \tag{A.3}$$

where R_t^A reflects the residual terms in equation (A.2), and R_t^B is the residual impact on sustainable income due to dynamic autonomous factors and revaluation effects. These new elements cause particular diffi-

---

[3]A thorough treatment of revaluation effects is provided by Aronsson et al. (1997).

culties for sustainability accounting because they generally cannot be easily measured or found in marketable assets or consumption.[4] In this most general expression, sustainable income is NDP plus the residual value provided by omitted consumption and capital formation plus the residual due to autonomous dynamic factors, plus the revaluation effects.

Departures from Efficient Decision Making

The present discussion of measurement of sustainable income depends on a stringent set of assumptions about the processes of social decision making. A set of qualifications concerns departures from the idealized assumptions of perfect competition, perfect foresight in asset markets, and an optimal allocation of resources over time. It is particularly important to understand that sustainable national income is equivalent to the maximum that can be consumed while leaving endowments sufficient to ensure equivalent living standards for the future. But to say that consumption *can* be sustained does not ensure that future living standards *will be* sustained. Consuming less than sustainable income provides the next generation the resources they need to keep the economy going at the same or improved living standards if they so choose. There is, however, no way to bind subsequent generations to serve as responsible trustees for their descendants. The current generation might leave a generous stock of oil, forests, and tangible capital, but some future generation might squander its inheritance through high living, bad judgments, or military adventures. But such behavior would not be reflected in current prices and quantities and therefore would not be reflected in current measures of sustainable income. This example illustrates one of the limitations of any income measure that relies on market prices and quantities.

EXAMPLE FOR PETROLEUM RESOURCES

The above approach can be illustrated with a specific example of how the sustainability framework can help in understanding real policy concerns. Consider the constraints on economic growth posed by the finiteness of petroleum stocks. What is the effect on sustainable income of the fact that we will run out of oil some day? Without an accounting frame-

---

[4]There is a rigorous derivation of the autonomous terms as the present value of the marginal contribution of the autonomous variables to current and future consumption. This contribution is not observable, however, and poses several forecasting and measurement problems.

work, it is difficult even to pose such a question meaningfully. Using the concept of sustainable income, it is possible to measure the impact of changing petroleum stocks on sustainable income by adjusting income with a new investment term that equals the market or scarcity price of petroleum reserves times the net change in those reserves. The fraction by which sustainability, and therefore future welfare, will be lowered as a result of running out of oil is in principle captured by the fraction of comprehensive product accounted for by the value of the change in net stocks of oil reserves. This measure, which is exactly the approach analyzed in Chapter 3, will provide a reasonably accurate estimate of the impact of depletions or additions on the sustainable consumption of the United States.

To better appreciate the power of these results, imagine for a moment that a dream team of world-class researchers is asked to analyze the important question of the impact on future living standards of the exhaustion of finite petroleum stocks. The researchers are provided with a large budget and told to make the best possible estimate. What would this dream research team do? They would project all relevant demand and supply response functions, estimate all sectoral rates of technological change, include all relevant elasticities of substitution between oil and everything else, project future technological advances, and so forth. They would then use thousands of linked parallel processors to simulate future trajectories with a massive, dynamic, computable general-equilibrium model. After this immense research project had been completed, the dream team would be asked to provide their best overall estimate of the impact of changing oil stocks on future living standards.

The surprising result is that estimates of the market value of oil depletions or additions probably offer the most accurate measure of the impact of changes in petroleum stocks on living standards. Moreover, the market's estimate is likely to be more credible than the dream team's estimate because it is based on an "invisible-hand" evaluation that is more reliable than the dream team's computer model. This invisible-hand model represents the judgment of the thousands of market participants who consider every relevant aspect of the problem treated by the dream team and additionally have their personal fortunes and livelihoods at risk if they make the wrong decisions.

SUMMARY

There are many different approaches to sustainability. From the point of view of a national economy and national income accounting, a useful definition is that sustainable national income is the maximum amount a nation can consume while ensuring that all future generations can have living standards at least as high as those of the current generation.

The NIPA have a close relationship to measures of sustainable income. The usual measure of NDP corresponds to the highest sustainable level of consumption under idealized conditions. The most important conditions underlying this correspondence are the inclusion of all consumption and net investment and the absence of technological change or other dynamic autonomous elements.

Measures of national income and output would be closer to the ideal measure of sustainable income if omitted consumption and net investment were included to obtain augmented income and output measures. Omitted items would include nonmarket consumption, such as home production, and final environmental services and nonmarket investment, such as changes in the value of resource stocks, along with investment in human capital.

When there is unmeasured consumption, investment, autonomous dynamic elements, or revaluation effects, residual terms must be added to reflect the contribution of these factors—positive or negative—to future income. Particularly important residual effects are due to technological change and institutional factors such as the nature of tangible and intellectual property rights and the degree of openness of the economy.

APPENDIX
B

Sources of Physical and Valuation Data on Natural Resources and the Environment

 Currently, substantial monitoring of physical flows and valuation of certain important resource and environmental assets and service flows within the United States are undertaken in conjunction with existing regulatory analysis and enforcement or as part of the activities of federal resource management agencies. These efforts include those of the Environmental Protection Agency, the National Oceanic and Atmospheric Administration, the National Center for Health Statistics, the Forest Service, the National Agricultural Statistical Service, the United States Geological Survey, the Fish and Wildlife Service, and other private and governmental organizations. For management and analysis purposes, valuation estimates have also been developed for some nonmarket goods and environmental effects. Table B-1 lists several existing studies of resource and recreational values for the United States, while Table B-2 provides important sources for natural-resource assets and recreational activity data.

TABLE B-1   Natural-Resource and Environmental Value Estimates

Source	Method/Unit/ Level of Detail	Activities Valued
Value of Human Health		
U.S. Environmental Protection Agency (1997)	21 labor market estimates and 5 contingent valuation studies	Mortality ($4.8 million per statistical life)
U.S. Environmental Protection Agency (1997)	Willingness-to-pay studies or cost of illness	Chronic bronchitis, ischemic heart disease, shortness of breath, acute bronchitis (ranges from $260,000 per case for chronic bronchitis to $5.30 per day for shortness of breath)
U.S. Environmental Protection Agency (1997)	Labor market studies	Work-loss days ($83 per day)
Surface Water Quality		
Ribaudo and Piper (1991)	Recreation demand model	National recreational fishing benefits from reduced sediment pollution
Carson and Mitchell (1993)	Contingent valuation method	Willingness to pay to improve the nation's water quality from nonboatable status to swimmable status
Wetlands		
Hoehn and Loomis (1993), Phillips et al. (1993), and Lant and Roberts (1990)	Contingent valuation method	Protection of wetlands and wildlife habitat and water quality, and quantity decrements to nonconverted wetlands
Groundwater Quality		
Sun et al. (1992)	Contingent valuation method	Keeping groundwater quality below EPA health advisory levels in southwestern Georgia
Jordan and Elnagheeb (1993)	Contingent valuation method	Protection from nitrate contamination of groundwater serving wells and drinking water utilities in Georgia

(continues)

TABLE B-1   Continued

Source	Method/Unit/ Level of Detail	Activities Valued
Recreation		
U.S. Fish and Wildlife Service (Hay, 1988)	Contingent valuation/user day/state by state	Deer, elk, and waterfowl hunting; bass fishing; and nonconsumptive uses
U.S. Fish and Wildlife Service (Waddington et al., 1994)	Contingent valuation/user day/state by state	Deer hunting, bass and trout fishing, and wildlife watching
U.S. Forest Service (McCollum et al., 1990)	Travel cost model/trips/ nine regions	Camping, swimming, hiking, viewing, hunting, picnicking, sightseeing, gathering products
U.S. Department of Agriculture (Ribaudo, 1989)	Travel cost model/trips/ten regions	Improvements to surface water use (fishing) from reductions in soil erosion
U.S. Forest Service (Sorg and Loomis, 1984)	Travel cost model and contingent valuation/ user day/selected states	Various activities, including fishing, hunting, camping, skiing, hiking, boating, picnicking, water sports, and nature viewing
Colorado Water Resources Research Institute (Walsh et al., 1988)	Travel cost model and contingent valuation/ user day/selected states	Various activities, including fishing, hunting, camping, skiing, hiking, boating, picnicking, water sports, and nature viewing
U.S. Forest Service (Bergstrom et al., 1996)	Travel cost model/user day/ten regions	Hiking, rafting, boating, cycling, picnicking, sightseeing, waterskiing, swimming, skiing, and hunting (20 activities total)
Biota		
Loomis and White (1996)	Contingent valuation/ animal species	Willingness to pay to protect 18 threatened and endangered species

TABLE B-2   Sources of Physical Data on the Environment and Natural
Resources

Resource/Source	Source/Comments
Timber Resource Planning Act (RPA)	Under the Resource Planning Act, the U.S. Department of Agriculture's Forest Service conducts renewable resource inventories of forest lands and collects statistics on forest products. These data are used to identify trends in extent, condition, ownership, quantity, and quality of timber and other forest resources.
Fish	The U.S. Department of Commerce's National Marine Fisheries Service in the National Oceanic and Atmospheric Administration collects and publishes data on the volume and value of commercial fish and shellfish landings, the catch by recreational fishermen, employment of people and craft in the fisheries, number of recreational fishermen, production of manufactured fishery products, and fishery prices.
Land Quality National Resources Inventory	The U.S. Department of Agriculture's Natural Resources Conservation Service conducts a survey every 5 years—the National Resources Inventory—to determine conditions and trends in the use of soil, water, and related resources nationwide and statewide. The National Resources Inventory is an inventory of land cover and use, soil erosion, prime farmland, wetlands, and other natural-resource characteristics on nonfederal rural land in the United States.
Air Quality National Air Quality and Emissions Trends Report	The Environmental Protection Agency (EPA) examines air pollution trends of each of the six principal pollutants in the United States. A yearly EPA document—the *National Air Quality and Emissions Trends Report*—gives an analysis of changes in air pollution levels plus a summary of current air pollution status.

Water Quality
National Water Quality
  Inventory Report to
  Congress
	The EPA publishes the *National Water Quality Inventory Report to Congress*, which summarizes water quality information submitted by 61
	(continues)

TABLE B-2   Continued

Resource	Source/Comments
	entities, including states, American Indian tribes, territories, interstate water commissions, and the District of Columbia. The report characterizes water quality in the United States, identifies widespread water quality problems of national significance, and describes various programs implemented to restore and protect water quality.
	The United States Geological Survey (USGS) collects information on the quality of ground and surface waters. During the past 30 years, the USGS has operated two national stream water quality networks—the Hydrologic Benchmark Network and the National Stream Quality Accounting Network. The data have been used to describe and quantify water quality trends.
Water Use	USGS compiles and publishes data on water use. The latest publication, *Estimated Use of Water in the United States in 1990,* describes water use by major water-use categories. For each category, there is a description of where the water comes from, what the water is used for, and where it goes after use.
Wetlands	The U.S. Department of the Interior's Fish and Wildlife Service conducts a National Wetlands Inventory to measure changes in wetlands. Changes are measured in acres and reported in *Status and Trends of Wetlands in the Conterminous United States,* required by Congress at 10-year intervals.
Biota	The National Biological Survey (NBS) researches and monitors trends in contaminant residue levels in birds and fish by geographic location. The NBS annual bird-banding program is conducted to determine the distribution, mortality, and survival of migratory game and nongame species. The U.S. Fish and Wildlife Service conducts an annual survey to monitor waterfowl, dove, and woodcock populations and waterfowl harvests.
Climate	The mission of the National Oceanic and Atmospheric Administration's National Climate Data Center is to manage global climatological data and information.

TABLE B-2   Continued

Resource	Source/Comments
Human Health	
National Center for Health Statistics	The National Center for Health Statistics (NCHS) is part of the Centers for Disease Control and Prevention, U.S. Department of Health and Human Services. NCHS is the federal government's principal vital and health statistics agency and provides a variety of data, including data on vital events, health status, lifestyle, exposure to unhealthy influences, onset and diagnosis of illness and disability, and use of health care.
Recreation	
The Fishing, Hunting, and Wildlife Associated Recreation Survey	National survey conducted every 5 years. It covers wildlife-associated recreation. It provides good information on private expenditures and mediocre information on site choice.
The National Survey of Recreation and the Environment	National survey conducted approximately every 10 years. A variety of outdoor recreational activities are covered. Some information on expenditures and site choice is provided.
State Comprehensive Outdoor Recreation Plan	State-specific surveys. Most states have such a plan since they are a prerequisite for receiving federal land acquisition monies. They are primarily an inventory of facilities and projected use levels and are of varied quality (the better ones contain county-level breakdowns).
Public Area Recreation Visitors Study/Customer Use and Survey Techniques for Operations, Management, Evaluation, and Research	Ongoing on-site surveys of U.S. Forest Service sites. Some information on expenditures is provided, along with good information on site choice.
The U.S. Forest Service's Recreation Reporting Information System	Yearly totals of the number of visitors to national forests. Some breakdown by activity is included.
Proprietary databases	A variety of subjects. For example, Hagler Baily Consultants has a database on water quality benefits.

APPENDIX
C

Accounting for Forest Assets

 While adjustment in an asset account is conceptually similar to net investment of "made assets," it is more precise for forests to call the change in asset values "net accumulation" to reflect the fact that, even at constant prices, the asset value of a forest can either increase or decrease. Most generally, net accumulation is defined[1] as

$$N(t) = V(t + 1) - V(t) \qquad (C.1)$$

where $N(t)$ is the net accumulation in period t, and $V(t)$ is the asset value in period t (present value of rents). As Hartwick and Hagemann (1993) have shown, net accumulation can also be written as:

$$N(t) = rV(t + 1)/(1 + r) - \{p\ q(t) - C[q(t)]\} \qquad (C.2)$$

where r = discount rate (or the difference between the nominal interest rate and the rate of growth of log prices, p is the price of logs, $q(t)$ is the harvest level in period t, and $C[q(t)]$ is total extraction costs. While these definitions are both general and precise, they are generally impossible to implement empirically: $V(t + 1)$ cannot be directly observed in period t.

---

[1]This development is done in discrete time to reflect (1) the annual growth period of temperate and boreal forests that characterizes virtually the entire United States, and (2) the annual reporting period that is recommended for forest accounting.

Instead, economists infer the value of the asset from assumptions about intertemporal market equilibrium. Three cases have been examined.

The first is identical to the literature on nonrenewable resources, appropriately treating the exploitation of primary, old-growth forests as timber mining. Since it is generally uneconomic to replace primary forests with forests of a similarly old age, this analogy is not as odd as it might seem. Under these circumstances,

$$N(t) = -\{p - C'[q(t)]\}q(t) \tag{C.3}$$

where $C'[q(t)]$ is marginal extraction costs. This model of net accumulation (pure depreciation) is generally called the Hotelling model to emphasize the connection between mining old growth that *will not* be replaced and mining minerals that *cannot* be replaced.

While the Hotelling model may be appropriate for the case of pure depreciation, it misses several important aspects of the forest sector. An alternative approach is transition models, which account in part for these problems by recognizing that forest growth offsets harvests. Assuming constant prices and a forest inventory recognized only by total net growth, this model suggests that net accumulation is

$$N(t) = (p - C')[g(t) - q(t)] \tag{C.4}$$

where $g(t)$ is the net forest growth in period t.

By recognizing forest growth, such a formulation improves on the ordinary Hotelling approach, but still suffers the defects of (1) ignoring endogenous price changes in the sector and (2) characterizing the forest only by net growth and not its more complex underlying age-class structure. Economic theory suggests that once the transition between old- and second-growth forests is complete, timber prices will stabilize and the economic return to holding forests will arise solely from forest growth. Vincent (1997) has developed the appropriate measures of net accumulation for optimally managed second-growth forests. Depreciation associated with the harvests equals

$$N_h(\tau^*) = -\{p - C'[v'(\tau^*)]\}v'(\tau^*)[1 - (1 + r)^{1-\tau^*}]/r \tag{C.5a}$$

where h is values per unit area, $v(\tau)$ is the timber yield at age τ, and τ^* is the economically optimal rotation age. Accumulation associated with the growth of subrotation-age forests is

$$N_h(\tau) = \{p - C'[v'(\tau^*)]\}v'(\tau^*)(1 + r)^{\tau - \tau^*} \tag{C.5b}$$

Net accumulation is simply the sum of unit-area depreciation or accumulation weighted by the area in the particular age class:

$$N(t) = \Sigma A(t,\tau) N_h(\tau) \text{ for } \tau = 1,....\tau^* \qquad (C.6)$$

where $A(t,\tau)$ is the area in age-class τ in period t. Note that if $A(t,\tau) = \Sigma A(\tau)/\tau^*$ for all τ (the so-called "normal forest" with an equal area in each age class), then $N(t) = 0$. This approach improves upon both the Hotelling and transition approaches. It assumes that forest owners cut their trees at the economically optimal time and that timber prices are in intertemporal market equilibrium.

The three cases discussed above require assumptions of intertemporal price equilibrium, optimal management, and constant prices and costs. These are strong assumptions. With the data that exist for U.S. forests, it is possible to develop a practical approach for measuring timber accumulation that improves upon the methods used in most countries and requires less restrictive assumptions.

Ideally, one would like transaction data on a representative sample of timberland. Because timberland is an extremely heterogenous product traded in dispersed markets, assembling such data is quite difficult.[2] It is therefore necessary to compute V(t) and V(t + 1) directly and to estimate net accumulation using equation (C.1). One possible approach, following Vincent (1997), begins by summing per-acre values weighted by area across all age classes (subscripts for other important value descriptors, such as region, ownership, species, and site quality, are suppressed for ease of reading).

$$V(t) = \Sigma A(t,\tau) V_h(t,\tau), \tau = 1,....T - 1 \qquad (C.7)$$

$$V_h(t,\tau) = [p(t)v(T) + p_\ell(t)] (1 + r)^{-(T - \tau)} \qquad (C.8)$$

$$p_\ell(t) = [-C_s(t) + p(t)v(T) (1 + r)^{-T}]/[1 - (1 + r)^{-T}] \qquad (C.9)$$

where T is actual cutting age, p(t) is net price (stumpage price), $p_\ell(t)$ is bare-land value, and $C_s(t)$ is full rotation management costs in period t.

---

[2]The National Council of Real Estate Fiduciaries maintains a database on the value of some industrial-grade timberland in three regions of the United States: the Pacific coast, the south, and the northeast. While these data are useful for measuring the return on timberland assets, they have severe limitations for present purposes. First, they reflect *industrial-grade* timberland, a category of timberland that probably covers no more than 20 percent of the total area of U.S. forestland. Second, the data generally do not reflect actual property transactions, but rather appraised values. Third, the number of properties in the sample, especially in the northeast, is small, and the data do not cover the north central or inland west regions at all.

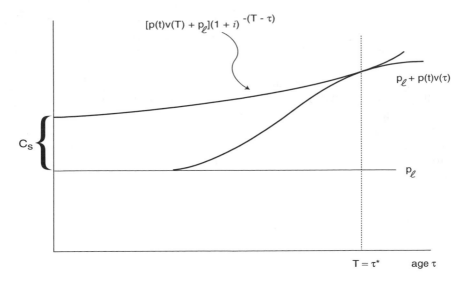

FIGURE C-1   Valuation of Forest Stands of Differing Ages.

The above are the basic valuation equations for forests of different age classes. Figure C-1 shows the approach graphically. The value of an acre of age-class τ timber is simply the value of the timber at the harvest age T (which may or may not be the economically optimal harvest age), discounted by the number of years until harvest. Note that p(t) in this formulation is net price, not price gross of extraction costs. This definition is used because the net price (called the stumpage price) is commonly recorded in forestry. While net prices differ across regions and among species, some evidence suggests that within a region, extraction costs may be constant (Adams, 1997). In this formulation, net rent = A(t,T)v(T)p(t) and is already accounted for in the National Income and Product Accounts for marketed timber. If desired, net accumulation can be divided into separate revaluation, growth, and depletion accounts in the usual way.

How would such an approach be implemented? The forest inventory and analysis work of the U.S. Department of Agriculture's Forest Service maintains data on A(t,τ), v(τ), T, and C_s(t) by eight regions in the United States, ownership, and site quality; unfortunately, the data on the national forests are not as comprehensive as those on private and other public lands. The Forest Service's ATLAS model is designed to update the inventory on an annual basis and to project it into the future. Data on p(t) by region and species (or species group) are available from the Forest

Service (the value of timber harvested in a year, or so-called "cut prices") and from commercial price reports for some regions.

The primary difficulty with this valuation approach is selecting the discount rate (r) to be used. Similar difficulties arise in minerals accounting when the net present value approach is introduced. No valuation method escapes this difficulty, except with strong and unrealistic assumptions about intertemporal price paths. A practical approach is to reestimate the discount rate periodically on the basis of an asset pricing model for timberland. The discount rate is then treated in the same manner as timber prices, with a separate revaluation account to reflect changes in the discount rate.

APPENDIX
D

Glossary

Accounts and accounting: The purpose of accounting is to provide economic information about a household, organization, or government. Accounts are generally divided into "income accounts," which record receipts and outlays during a given period such as a year, and "asset accounts," which provide a snapshot of the assets, liabilities, and net worth of an entity at a given date. People are most familiar with the income accounts and balance sheets of businesses, but the same concepts apply equally well to individuals, governments, and nations.

Air pollutants: Substances in the air that could, at high enough concentrations, harm human beings, animals, vegetation, or material. Air pollutants may thus include forms of matter of almost any natural or artificial composition capable of being airborne. They may consist of solid particles, liquid droplets, or combinations of these.

Air quality standards: Levels of air pollutants, prescribed by regulations, that may not be exceeded during a specified time in a defined area.

Ambient concentration: Measure of environmental quality indicating the amount of pollutants found per unit volume in different environmental media.

Assimilation: Ability of natural systems to safely absorb waste and residuals.

Associated capital: The capital investments attached to a resource that is accounted for elsewhere in the core national accounts. Examples include machinery, the capitalized value of mineral exploration, and access roads.

Augmented accounts: See *satellite accounts*.

207

Avoidance costs: Actual or imputed costs for preventing environmental deterioration by alternative production and consumption processes, or by the reduction of or abstention from economic activities.

Background concentration: Ambient concentration of pollutants, such as carbon dioxide and other greenhouse gases, measured by background stations.

BEA: See *Bureau of Economic Analysis.*

Biodiversity: Range of genetic differences, species differences, and ecosystem differences in a given area.

Biomass: Total living weight (generally in dry weight) of all organisms in a particular area or habitat. It is sometimes expressed as weight per unit area of land or per unit volume of water.

Biosphere: Thin stratum of the earth's surface and upper water layer containing the total mass of living organisms that process and recycle the energy and nutrients available from the environment.

Bureau of Economic Analysis (BEA): An agency of the U.S. Department of Commerce that serves as the nation's economic accountant, preparing estimates that illuminate key national, international, and regional aspects of the U.S. economy.

Capital: In classical and neoclassical economic theory, one of the triad of productive inputs (land, labor, capital). Capital consists of durable produced goods that are in turn used in production. The major components of capital are equipment, structures, and inventory.

Capital accumulation (environmental accounting): Environmentally adjusted concept of capital formation that accounts for additions to and subtractions from natural capital. The concept may also include discoveries or transfers (from the environment into the economic system) of natural resources, and the effects of disasters and natural growth.

Capital consumption: The wearing away of capital stock due to physical destruction or erosion through the ravages of time and through the use of the asset in production, plus the complete withdrawal of capital assets from capital stock (scrappage). Depreciation is more general, in that it is the fall in the price of a capital asset as it ages. Depreciation includes capital consumption, and it also includes revaluation, which consists of pure inflation and obsolescence.

Carbon dioxide (CO_2): Colorless, odorless, and nonpoisonous gas that results from fossil fuel combustion and is normally a part of ambient air. It is also produced in the respiration of living organisms (plants and animals) and considered to be the main greenhouse gas contributing to climate change.

Carbon monoxide (CO): Colorless, odorless, and poisonous gas produced by incomplete fossil fuel combustion. Carbon monoxide combines with the hemoglobin of human beings, reducing the latter's oxygen-carrying capacity, with effects harmful to human health.

Carbon sink: Pool (reservoir) that absorbs or takes up released carbon from another part of the carbon cycle. For example, if the net exchange between the biosphere and the atmosphere is toward the atmosphere, the biosphere is the source, and the atmosphere is the sink.

Carbon tax: Instrument of environmental cost internalization. It is a tax on the producers or users of raw fossil fuels, based on the relative carbon content of those fuels.

Chlorofluorocarbons (CFCs): Inert, nontoxic, and easily liquefied chemicals used in refrigeration, air conditioning, packaging, and insulation, or as solvents and aerosol propellants. Because CFCs are not destroyed in the lower atmosphere, they drift into the upper atmosphere, where their chlorine components destroy ozone. They are also among the greenhouse gases that affect climate change.

Climate change: Term frequently used in reference to global warming due to greenhouse gas emissions from human activities. See also *greenhouse effect*.

Commission of the European Communities: Revised (1993) system adopted worldwide for conventional economic (national) accounting (Commission of the European Communities et al., 1993).

Complement: A relationship between goods or services in which a rise in the price of one decreases demand for the other.

Consumer surplus: Difference between the amount a consumer would be willing to pay for a commodity and the amount he or she actually pays.

Consumption: Total spending, by individuals or a nation, on consumer goods during a given period.

Contaminant: Any physical, chemical, biological, or radiological substance or matter that has an adverse effect on air, water, land/soil, or biota. The term is frequently used synonymously with *pollutant*.

Contingent valuation: Method of valuation used in cost-benefit analysis and environmental accounting. It is conditional (contingent) on the construction of hypothetical markets, and is one method of estimating the willingness to pay for potential environmental benefits or for the avoidance of their loss.

Core accounts: National Income and Product Accounts or traditionally and regularly reported accounts leading to such overall measures as *gross domestic product (GDP)*.

Cost: Measure of what must be given up to acquire or achieve something.

Cost-benefit analysis: Assessment of the direct economic and social costs and benefits of a proposed program for the purpose of program selection. The cost-benefit ratio is determined by dividing the projected benefits of the program by the projected costs.

Cost internalization: Incorporation of negative external effects, notably environmental depletion and degradation, into the budgets of households and enterprises by means of economic instruments, including fiscal measures and other (dis)incentives.

Current rent method: Valuation method that relies largely on the current rents or economic profits from harvesting or extraction.

CV: See *contingent valuation.*

Defensive environmental costs: Actual environmental protection costs incurred in preventing or neutralizing a decrease in environmental quality, as well as the expenditures necessary to compensate for or repair the negative effects (damage) of environmental deterioration. Such costs include expenditures required to mitigate environment-related health and other welfare effects on human beings.

Depletion costs: Monetary value of the quantitative depletion (beyond replenishment or regeneration) of natural assets by economic activities. Depletion of natural resources results from their use as raw materials in production or directly in final (household) consumption.

Development: Process of making economic resources available or useful.

Discounting (of natural assets): Determining the present value (net worth) of assets by applying a discount rate to the expected net benefits from future uses of those assets. The discount rate reflects the social preferences for current (as compared with future) uses.

Earth Summit: See *United Nations Conference on Environment and Development.*

Ecological impact: Effects of human activities and natural events on living organisms and their nonliving environment. See also *environmental impact.*

Economic activity: The production, consumption, or transformation of value or utility.

Economic assets: Assets recorded in the balance sheets of conventional national accounts. Economic assets are entities (1) over which ownership rights are enforced by institutional units, individually or collectively, and (2) from which owners may derive economic benefits by holding or using the asset over a period of time.

Economic instruments: Fiscal and other economic incentives and disincentives to incorporate environmental costs and benefits into the budgets of households and enterprises. The objective is to encourage environmentally sound and efficient production and consumption through full-cost pricing. Economic instruments include effluent taxes and charges on pollutants and waste, deposit-refund systems, and tradable pollution permits. See also *cost internalization*.

Economic profit: See *rent*.

Economic rent: See *rent*.

Emission factor: Ratio between the amount of pollution generated and the amount of a given raw material processed. The term may also refer to the ratio between the emissions generated and the outputs of production processes.

Emission standard: Maximum amount of polluting discharge legally allowed from a single source, mobile or stationary.

Energy Information Administration (EIA): An agency of the U.S. Department of Energy that collects and analyzes energy related data.

Environmental accounting: In national accounting, physical and monetary accounts of environmental assets and the costs of their depletion and degradation. In corporate accounting, the term usually refers to environmental auditing, but may also include the costing of environmental impacts caused by the corporation. See also *System of Integrated Environmental and Economic Accounting (SEEA)*.

Environmental and Natural Resource Accounting Program (ENRAP): Philippine program of environmental and natural-resource accounting.

Environmental assets: See *natural assets*.

Environmental cleanup: Action taken to deal with the release of a hazardous substance that could affect humans and/or the environment. The term is sometimes used interchangeably with the terms "remedial action," "response action," and "corrective action," as opposed to the terms "preventive action" and "anticipatory action." See also *environmental restoration* and *environmental protection*.

Environmental costs: Costs connected with the actual or potential deterioration of natural assets due to economic, social, or political activities. Such costs can be viewed from two different perspectives: (1) as costs caused, that is, costs associated with economic units actually or potentially causing environmental deterioration by their own activities, or (2) as costs borne, that is, costs incurred by economic units independently of whether they have actually caused the environmental impacts. See also *defensive environmental costs*.

Environmental damages: Harm caused to the environment by natural or human activities. They are frequently measured in dollars, but some damages may be unmeasurable.

Environmental degradation: Deterioration in environmental quality from ambient concentrations of pollutants and other activities and processes, such as improper land use and natural disasters.

Environmental effect: Result of an impact of the environment on human health and welfare.

Environmental expenditures: Capital and current expenditures related to characteristic activities and facilities specified in classifications of environmental protection activities.

Environmental externalities: Economic concept of uncompensated environmental effects of production and consumption that affect consumer utility and enterprise cost outside the market mechanism. As a consequence of negative externalities, private costs of production tend to be lower than "social" costs. It is the aim of the "polluter/user pays" principle to prompt households and enterprises to internalize externalities in their plans and budgets. See also *economic instruments*.

Environmental functions: Environmental services, including spatial functions, waste disposal, natural resource supply, and life support. See also *environmental services*.

Environmental impact: Direct effect of socioeconomic activities and natural events on the components of the environment. See also *environmental effect*.

Environmental protection: Any activity to maintain or restore the quality of environmental media by preventing the emission of pollutants or reducing the presence of polluting substances. It may consist of (1) changes in characteristics of goods and services, (2) changes in consumption patterns, (3) changes in production techniques, (4) treatment or disposal of residuals in separate environmental protection facilities, (5) recycling, and (6) prevention of degradation of the landscape and ecosystems.

Environmental Protection Agency (EPA): A federal agency of Cabinet rank. The EPA obtains and analyzes information pertaining to the environment and suggests and enforces relevant federal laws designed to protect human health and the natural environment.

Environmental protection costs: Costs associated with preventing environmental damage.

Environmental quality standard: Limit for environmental disturbances, in particular from ambient concentrations of pollutants and wastes, that determines the maximum allowable degradation of environmental media.

Environmental restoration: Reactive environmental protection. It includes: (1) reduction or neutralization of residuals; (2) changes in the spatial distribution of residuals; (3) support for environmental assimilation; and (4) restoration of ecosystems, landscape, and so forth. See also *environmental protection*.

Environmental services: Qualitative functions of natural nonproduced assets of land, water, and air (including related ecosystems) and their biota. There are three basic types of environmental services: (1) disposal services, which reflect the functions of the natural environment as an absorptive sink for residuals; (2) productive services, which reflect the economic functions of providing natural-resource inputs and space for production and consumption; and (3) consumer or consumption services, which provide for physiological as well as recreational and related needs of human beings.

Environmental statistics: Statistics that describe the state and trends of the environment, covering the media of the natural environment (air/climate, water, land/soil). Environmental statistics are integrative in nature, measuring human activities and natural events that affect the environment, the impacts of these activities and events, social responses to environmental impacts, and the quality and availability of natural assets. Broad definitions include environmental indicators, indices, and accounting.

European System for the Collection of Economic Information on the Environment (SERIEE): System consisting mainly of data on environmental protection expenditures and economic data on the use and management of natural resources. Links to physical data, such as the amount of waste and other pollutants generated or avoided and the use of water and other resources, are to be established in parallel as far as possible. The system is designed to form a series of satellite accounts of the national accounts.

Existence value: Value of knowing that a particular species, habitat, or ecosystem does and will continue to exist. Such value is independent of any use the valuer may or may not make of the resource.

Externality: Activity that affects others for better or worse, without those others paying or being compensated for the activity. Externalities exist when private costs or benefits do not equal social costs or benefits.

FASAB: Federal Accounting Standards Advisory Board. Considers and recommends accounting principles for the federal government. Established in October 1990 by the secretary of the Department of the Treasury, the director of the Office of Management and Budget, and the comptroller general of the United States. It is an advisory committee operating under the Federal Advisory Committee Act.

Fixed capital: Traditionally nonresidential structures, residential structures, and producers' durable equipment. Extended accounts would also include environmental and human assets.

Flow vs. stock: A flow variable is one that has a time dimension or that flows over time (like a stream); a stock variable is one that measures a quantity at a point in time (like the water in a lake).

Fossil fuels: Coal, oil, peat, and natural gas.

Fugitive assets: Assets capable of moving of their own accord, e.g., wild or natural fish.

GDP: See *gross domestic product.*

GNP: See *gross national product.*

Genetic resources: Genetic material of plants, animals, or microorganisms of value as a resource for future generations of humanity.

Global warming: Phenomenon believed to occur as a result of the buildup of carbon dioxide and other greenhouse gases, identified by many scientists as a major global environmental threat. See also *greenhouse effect.*

Green GDP: Popular term for environmentally adjusted gross domestic product.

Greenhouse effect: Warming of the earth's atmosphere caused by a buildup of carbon dioxide and other greenhouse or trace gases. These gases allow sunlight to pass through and heat the earth, but prevent a counterbalancing loss of heat radiation. See also *global warming.*

Greenhouse gases: Carbon dioxide, nitrous oxide, methane, ozone, and chlorofluorocarbons occurring naturally or resulting from human (production and consumption) activities and contributing to global warming. See also *global warming* and *greenhouse effect.*

Gross domestic product (GDP): The most important item in the National Income and Product Accounts (NIPA). GDP measures the nation's total output of goods and services and the total income of the nation generated by that output. It measures the sum of the dollar values of consumption, gross investment, government purchases of goods and services, and net exports produced within a nation during a given year, where these transactions are valued at market prices. It also represents the incomes earned as wages, profits, and interest, as well as indirect taxes. In addition to the totals for the nation, the NIPA provide a rich array of data on output and incomes in different industries and regions, as well as a record of international transactions.

Gross national product (GNP): The value at current market prices of all final goods and services produced during a period by the assets owned by the citizens of a nation.

Hedonic modeling: Running a regression analysis to measure or estimate the relationship between price and the characteristics of goods and services.

Hedonic pricing: Price as a function of the characteristics of a good or service; for example, the more skilled a laborer, the higher the price of his/her labor.

Hicksian income: The maximum a nation can consume while keeping its capital stock intact.

Hotelling rent: Net return realized from the sale of a natural resource under particular conditions of long-term market equilibrium. It is defined as the revenue received minus all marginal costs of resource exploitation, exploration, and development, including a normal return to fixed capital employed. The Hotelling rent is used as a measure of natural-resource depletion in environmental accounting.

Hotelling valuation: Valuing "mineable" natural resources using only current rents, avoiding the need to forecast future revenues and to discount them at some discount rate.

Human capital: Productive wealth embodied in labor, skills, and knowledge.

Inflow: Entry of extraneous rainwater into a sewer system from sources other than infiltration, such as basement drains, manholes, storm drains, and street washing.

Land degradation: Reduction or loss of the biological or economic productivity and complexity of rain-fed crop land, irrigated crop land, range, pasture, forest, or woodlands resulting from natural processes; land uses; or other human activities and habitation patterns, such as land contamination, soil erosion, and destruction of the vegetation cover.

Land reclamation: Gain of land from the sea, wetlands, or other water bodies and restoration of productivity or use to lands that have been degraded by human activities or impaired by natural phenomena.

Land-use classification: Classification providing information on land cover and the types of human activity involved in land use. It may also facilitate the assessment of environmental impacts on, and potential or alternative uses of, land. The classification, developed by the Economic Commission for Europe, consists of seven main categories: (1) agricultural land; (2) forest and other wooded land; (3) built-up and related land, excluding scattered farm buildings; (4) wet open land; (5) dry open land with special vegetation cover; (6) open land without or with insignificant vegetation cover; and (7) waters.

Loading: The quantity of polluting material discharged into a body of water.

Maintenance (cost) valuation (environmental accounting): Method of measuring imputed environmental (depletion and degradation) costs caused by economic activities of households and industries. The value of the maintenance cost depends on the avoidance, restoration, replacement, or prevention activities chosen.

Marginal cost: Increase in total cost required to produce 1 extra unit of output (or reduction in total cost from producing 1 less unit).

Marginal value: Dollar value of one additional unit of product.

Market valuation: (1) Market price valuation applied in national accounts; (2) value of natural resources and of their depletion and degradation, imputed in environmental accounting and estimated on the basis of expected market returns. See also *discounting (of natural assets)* and *Hotelling rent.*

Materials and energy balances: Accounting tables that provide information on the material input into an economy delivered by the natural environment, the transformation and use of that input in economic processes (extraction, conversion, manufacturing, consumption), and its return to the natural environment as residuals (wastes). The accounting concepts involved are founded on the first law of thermodynamics, which states that matter (mass/energy) is neither created nor destroyed by any physical process.

Maximum sustainable yield: Maximum use a renewable resource can sustain without its renewability being impaired through natural growth or replenishment.

McKelvey box: Two-dimensional scheme that combines criteria of increasing geologic assurance (undiscovered/possible/probable/proved reserves) with those of increasing economic feasibility (subeconomic "resources" as compared with economic "reserves," depending on price and cost levels).

Measure of economic welfare: Adjusted measure of total national output, including only the consumption and investment items that contribute directly to economic well-being. It is calculated as additions to gross national product (GNP), including the value of leisure and the underground economy, and deductions such as environmental damage. It is also known as "net economic welfare."

Mineral: A naturally occurring inorganic substance having a characteristic set of physical properties, a definite range of chemical composition, and a molecular structure usually in crystalline form.

Mineral reserve: Mineral resources that are both currently profitable to exploit and known with considerable certainty.

Mineral resource: A concentration of naturally occurring solid, liquid, or gaseous material in or on the earth's crust in such form and amount that economic extraction of a commodity from the concentration is currently or potentially feasible.

Mining wastes: Mining-related by-products of two types: (1) mining and quarrying extraction wastes, which are barren soils removed from mining and quarrying sites during the preparation for mining and quarrying that do not enter into the dressing and beneficiating processes, and (2) mining and quarrying dressing and beneficiating wastes, which are obtained during the process of separating minerals from ores and other materials extracted during mining and quarrying activities. These wastes occupy valuable land and cause harm to stream life when they are deposited near the drainage area of a stream.

National Income and Product Accounts (NIPA): Provide a coherent and comprehensive picture of the nation's economy. These accounts measure the total income and output of the entire nation, including households, businesses, not-for-profit enterprises, and different levels of government. The key concepts in the NIPA, the *core accounts*, measure the total market output and income of the United States (see Young and Tice, 1985; and Kendrick, 1996). For the most important item, see *gross domestic product (GDP)*.

Natural assets: Assets of the natural environment. They consist of biological assets (produced or wild), land and water areas with their ecosystems, subsoil assets, and air.

Natural resources: Natural assets that can be used for economic production or consumption. See also *renewable natural resources* and *nonrenewable natural resources*.

NDP: See *net domestic product*.

Near market: Goods or services that are provided outside of a market are called near-market goods or services. Goods or services can be provided both via markets and outside of markets. An example is vegetables that are purchased in a store (or market) versus those grown for home consumption.

Net domestic product (NDP): Gross domestic product less the allowance for depreciation of capital goods.

Net national welfare: See *measure of economic welfare*.

Net present value: Present value of an investment, found by discounting all current and future streams of income by an appropriate rate of interest.

Net present worth: Net present value of an organization's assets (after deduction of liabilities), calculated by discounting current and future streams of income by the appropriate interest rate.

Net price: Valuation used in natural-resource economics to estimate the economic value of a natural resource and its depletion. It is defined as the actual market price of a natural resource output minus all marginal exploitation costs, including a normal return to capital.

NIPA: See *National Income and Product Accounts.*

Nonmarket: Economic activity that produces goods and services not distributed by markets.

Nonrenewable natural resources: Exhaustible natural resources, such as mineral resources, that cannot be regenerated after exploitation.

NPV: See *net present value.*

OECD: Organization for Economic Cooperation and Development.

Opportunity cost: Value of the next best use (or opportunity) for an economic good, or value of the sacrificed alternative.

Ozone depletion: Destruction of ozone in the stratosphere, where it shields the earth from harmful ultraviolet radiation. Its destruction is caused by chemical reactions in which oxides of hydrogen, nitrogen, chlorine, and bromine act as catalysts.

Particulate loadings: Mass of particles per unit volume of air or water.

Particulate removal: Removal of particulate air pollutants from their gaseous media using gravitational, centrifugal, electrostatic and magnetic forces, thermal diffusion, or other techniques.

Particulates: Fine liquid or solid particles, such as dust, smoke, mist, fumes, or smog, found in air or emissions.

Photochemical air pollution: Pollution caused by the reaction of unsaturated and saturated hydrocarbons, aromatics, and aldehydes (emitted owing to the incomplete combustion of fuels) with light. It causes eye irritation.

Physical accounting: Natural-resource and environmental accounting of stocks and changes in stocks in physical (nonmonetary) units, for example, weight, area, or number. Qualitative measures, expressed in terms of quality classes, types of uses, or ecosystem characteristics, may supplement quantitative measures. The combined changes in asset quality and quantity are called "volume changes."

Pigouvian tax: Tax levied on an agent causing an environmental externality (environmental damage) as an incentive to avert or mitigate such damage.

Pollutant: Substance present in concentrations that may harm organisms (humans, plants, and animals) or exceed an environmental quality standard. The term is frequently used synonymously with *contaminant.*

Pollution: (1) Presence of substances and heat in environmental media (air, water, land) whose nature, location, or quantity produces undesirable environmental effects; (2) activity that generates pollutants.

Pollution abatement: Technology applied or measure taken to reduce pollution and/or its impacts on the environment. The most commonly used technologies are scrubbers, noise mufflers, filters, incinerators, wastewater treatment facilities, and comporting of wastes.

Pollution abatement costs or expenditures: Costs incurred to reduce or mitigate specific pollution.

Primary energy consumption: Direct use at the source, or supply to users without transformation, of crude energy, that is, energy that has not been subjected to any conversion or transformation process.

Proved reserves: Such estimated quantities of mineral deposits, at a specific date, as analysis of geologic engineering data demonstrates with reasonable certainty to be recoverable under the current economic and operational conditions, even though actual extraction may occur in the future.

Public good: A commodity whose benefits may be provided to all people (in a nation or town) at no more cost than that required to provide it for one person. The benefits of the good are indivisible, and people cannot be excluded from using it.

Public investment: Government spending on public goods.

Quality of life: Notion of human welfare (well-being) measured by social indicators rather than by quantitative measures of income and production.

Recreational land: Land used for purposes of recreation, for example, sports fields, gymnasiums, playgrounds, public parks and green areas, public beaches and swimming pools, and camping sites.

Renewable energy sources: Energy sources including solar energy, geothermal energy, wind power, hydropower, ocean energy (thermal gradient, wave power, and tidal power), biomass, animal power, and fuel wood.

Renewable natural resources: Natural resources that, after exploitation, can return to their previous stock levels by natural processes of growth or replenishment. "Conditionally renewable resources" are those whose exploitation eventually reaches a level beyond which regeneration becomes impossible. Such is the case with the clear-cutting of tropical forests.

Rent: Net return on a production factor whose supply is perfectly inelastic (available only as a fixed amount), such as land. It is also called "pure economic rent." See also *Hotelling rent*.

Restoration costs: Actual and imputed expenditures for activities aimed at the restoration of depleted or degraded natural systems, partly or completely counteracting the (accumulated) environmental impacts of economic activities. See also *environmental restoration*.

Ricardian rent: Any return to a factor of production fixed in supply.

Rio Declaration on Environment and Development: See *United Nations Conference on Environment and Development* (United Nations, 1993).

Risk assessment: Quantitative and qualitative evaluation of the risk posed to human health and/or the environment by the actual or potential presence of and exposure to particular pollutants.

Risk management: Process of evaluating and selecting among alternative regulatory and nonregulatory responses to risk. The selection process necessarily requires consideration of legal, economic, and social factors.

Royalty: Payment for the use of assets, both intangible, such as patents, and tangible, notably subsoil assets. Royalties paid for the use of subsoil assets are also called *rents* even though they are not rents by the definition given above.

Satellite accounts: Additional or parallel accounting system that expands the analytical capacity of national accounts without overburdening or disrupting the central system. It may provide additional information, apply complementary or alternative concepts, extend the coverage of costs and benefits of human activities, and link physical with monetary data. The *System of Integrated Environmental and Economic Accounting (SEEA)* constitutes a satellite system of the *System of National Accounts (SNA)*.

Secondary air pollution: Pollution caused by reactions in air already polluted by primary emissions (from factories, automobiles, and so forth). An example of secondary air pollution is photochemical smog.

Secondary treatment: Second step in most waste treatment systems, during which bacteria consume the organic portions of the wastes. This is accomplished by bringing the sewage, bacteria, and oxygen together in trickling filters or within an activated sludge process. Secondary treatment removes all floating and settleable solids and about 90 percent of oxygen-demanding substances and suspended solids. Disinfection by chlorination is the final stage of the secondary treatment process.

SEEA: See *System of Integrated Environmental and Economic Accounting*.

SNA: See *System of National Accounts*.

Strip mining: Process in which rock and topsoil strata overlying mineral deposits are removed in strips.

Stumpage value: Economic value of a standing tree, equivalent to the amount concessionaires earn when a log is sold to the sawmill or the exporter, less the cost of logging. It is used as the net-price valuation in environmental accounting.

Subsoil assets: Developed and undeveloped reserves of mineral deposits located on or below the earth's surface.

Supplemental accounts: See *satellite accounts*.

Sustainability: (1) Use of the biosphere by present generations while maintaining its potential yield (benefit) for future generations; (2) nondeclining trends of economic growth and development that might be impaired by natural-resource depletion and environmental degradation.

Sustainable development: "Development that meets the needs of the present generation without compromising the ability of future generations to meet their own needs" (World Commission on Environment and Development, 1987). It assumes the conservation of natural assets for future growth and development.

Sustainable income: Sustainable national income is defined as the maximum amount a nation can consume while ensuring that future generations will have living standards at least as high as those of the current generation.

System of Integrated Environmental and Economic Accounting (SEEA): Satellite system of the System of National Accounts (SNA) proposed by the United Nations (1993) for the incorporation of environmental concerns (environmental costs, benefits, and assets) into national accounts.

System of National Accounts (SNA): See Commission of the European Communities.

Tangible assets: Assets including human-made (produced) nonfinancial assets and nonproduced natural assets and excluding intangible (nonproduced) assets such as patents or good will. See also *natural assets*.

Technological change: Improvement in technology that allows for more output created by the same amount of inputs.

Tradable pollution permits: Rights to sell and buy actual or potential pollution in artificially created markets. See also *economic instruments*.

Transboundary pollution: Pollution that originates in one country but, by crossing the border through pathways of water or air, is able to cause damage to the environment in another country.

Travel cost: A method for assessing willingness to pay using cost data associated with movement to an environmental recreation area.

United Nations Conference on Environment and Development: Conference held in 1992 in Rio de Janeiro (also referred to as the Biodiversity Convention, the Climate Convention, and the Earth Summit). The conference adopted the Rio Declaration on Environment and Development; an action plan termed Agenda 21; and the Non-Legally Binding Authoritative Statement of Principles for a Global Consensus on the Management, Conservation and Sustainable Development of All Types of Forests (Forest Principles) (United Nations, 1993). The conference also presented for signature by governments the United Nations Framework Convention on Climate Change (United Nations, 1992a) and the Convention on Biological Diversity (United Nations Environment Program, 1992b).

United Nations Environment Program: International organization established in 1972 to catalyze and coordinate activities aimed at increasing scientific understanding of environmental change and developing environmental management tools.

User cost: Concept proposed for valuation of the depletion of mineral deposits (El Serafy, 1989). A time-bound stream of net revenues from the sale of an exhaustible natural resource is converted into a permanent income stream by investing part of the revenues—the user cost allowance—over the lifetime of the resource. The remaining amount of the revenue is regarded as true income.

Utility: The total satisfaction derived from the consumption of goods or services.

Valuation of natural assets: Methods of applying a monetary value to natural assets in environmental accounting that include (1) market valuation; (2) direct nonmarket valuation, such as assessment of the willingness to pay for environmental services (contingent valuation); and (3) indirect nonmarket valuation, for example, costing of environmental damage or of compliance with environmental standards. See also *maintenance (cost) valuation, market valuation,* and *contingent valuation.*

Value added: Difference between the value of goods produced and the cost of materials and supplies used in producing them.

Waste: Materials that are not prime products (that is, products produced for the market), for which the generator has no further use in terms of his/her own purposes of production, transformation, or consumption, and of which he/she wishes to dispose. Wastes may be generated during the extraction of raw materials, the processing of raw materials into intermediate and final products, the consumption of final products, and other human activities. Residuals recycled or reused at the place of generation are excluded.

Water conservation: Preservation, control, and development of water resources, both surface and groundwater, and prevention of pollution.

Water quality index: Weighted average of selected ambient concentrations of pollutants, usually linked to water quality classes.

Willingness to pay: See *contingent valuation.*

WRI: World Resources Institute.

APPENDIX
E

Biographical Sketches

WILLIAM D. NORDHAUS *(Chair)* is A. Whitney Griswold Professor of Economics at Yale University. He is on the staff of the Cowles Foundation and is a Research Associate of the National Bureau of Economic Research. He is a member of the Committee on National Statistics and has served on several National Research Council panels, including the Committee on Carbon Dioxide Assessment (Commission on Geosciences, Environment, and Resources), the Committee on Alternative Energy Research and Development Strategies (Commission on Engineering and Technical Systems), the Policy Implications of Greenhouse Warming Synthesis Panel and Adaptation Subpanel (Policy Division), and the Committee on the Human Dimensions of Global Change (Policy Division, Commission on Behavioral and Social Sciences and Education). He is a former member of the President's Council of Economic Advisers, is a senior advisor for the Brookings Institution Panel on Economic Activity, and served as Provost of Yale University from 1986 to 1988.

CLARK S. BINKLEY is Senior Vice President, Investment Strategy and Research, for the Hancock Timber Resource Group. He was formerly Dean of the Faculty of Forestry and Professor of Forest Resources Management at the University of British Columbia during the panel's deliberations. He holds a Ph.D. in forestry and environmental studies from Yale University, a master's degree in engineering from Harvard University, and an AM degree in applied mathematics from Harvard University. He has consulted with and acted as a member of several governmental and private forest products groups. He is an expert in timberland valuation and has written on carbon sequesterization, forest policy, and international forest prices.

ANU DAS was formerly Research Assistant with the Committee on National Statistics. In addition to the Panel on Integrated Environmental and Economic Accounting, she worked with the Panel on Statistical Methods for Testing and Evaluating Defense Systems and with the Longitudinal Research on Children Workshop. She previously worked on studies related to health, aging, disability, and census. She holds a bachelor's degree in mathematics and a master's degree in computer and information science. She currently works for the International Monetary Fund.

GRAHAM DAVIS is Assistant Professor of Mineral Economics at the Colorado School of Mines. His research includes the valuation of mineral assets, and he is the recipient of an Environmental Protection Agency/ National Science Foundation grant to investigate methods of incorporating these valuations into "green" national income accounting exercises. Before his academic career, he worked as a metallurgical engineer at mines in Namibia and Canada. He has an MBA in finance from the University of Cape Town and a Ph.D. in mineral economics from Pennsylvania State University. In 1996, Pennsylvania State University's College of Earth and Mineral Sciences elected him Centennial Fellow for his contributions to the field.

JOSHUA S. DICK is Senior Project Assistant with the Committee on National Statistics. In addition to working with the Panel on Integrated Environmental and Economic Accounting, he works with the Panel To Study the Research Program of the Economic Research Service (ERS), a Study To Review the Statistical Procedures for the Decennial Census, and a Study on Conceptual, Measurement, and Other Statistical Issues in Developing Cost-of-Living Indexes for Indexing Federal Programs. He holds a bachelor's degree in political science with honors from Florida Atlantic University and served as an intern for U.S. Senator Connie Mack. He is a member of Pi Sigma Alpha, the national political science honor society, and is an Eagle Scout with the Boy Scouts of America.

ROBERT EISNER was William R. Kenan Professor of Economics, Emeritus, at Northwestern University. He was a past president of the American Economic Association and a fellow of the American Academy of Arts and Sciences and the Econometric Society. He worked to extend the conventional National Income and Product Accounts and develop more comprehensive measures of output, market and nonmarket activity, investment, and the products of governments and households.

PETER M. FEATHER is employed by the U.S. Department of Agriculture's Economic Research Service as a research economist. He received his Ph.D. in applied economics in 1992 from the University of Minnesota. His research interests include nonmarket valuation methods, welfare measurement, and labor economics.

DANIEL HELLERSTEIN is Natural Resource Economist for the Economic Research Service of the U.S. Department of Agriculture. He received his Ph.D. from the Yale School of Forestry and Environmental Studies in 1989, He has published widely in the environmental economics literature, with a primary focus on the valuation of environmental benefits using travel-cost models. He is also the author of several publicly available software suites, including modeling packages used in the construction and econometric analysis of travel cost and other valuation models, Internet server software, and other utilities.

JAMES HRUBOVCAK is Economist with the Economic Research Service of the U.S. Department of Agriculture. He has conducted research in the areas of agricultural sustainability and environmental accounting. In addition, he has investigated factors affecting the capital structure of and investment in agriculture, with emphasis on the economics of taxation and the farm sector. He received his Ph.D. in economics from The George Washington University.

DALE JORGENSON is Frederic Eaton Abbe Professor of Economics and Chairman of the Economics Department at Harvard University. He received his Ph.D. in economics from Harvard. He was elected to membership in the National Academy of Sciences in 1978. In addition, he is a member of the American Academy of Arts and Sciences, the American Philosophical Society, and the Royal Swedish Academy of Sciences and a fellow of the American Association for the Advancement of Science. The American Economic Association presented him with the John Bates Clark Medal for his excellence in economic research.

EDWARD C. KOKKELENBERG, study director, is Associate Professor and Chairman of the Department of Economics, Binghamton University (SUNY). He joined Binghamton in 1980 after earning his Ph.D. in economics from Northwestern University. His teaching and publications have focused on energy, forecasting, the demand for capital and labor, productivity, and capacity utilization. He is a member of the American

Economic Association and the American Statistical Association and a Census-National Science Foundation Fellow.

BRIAN NEWSON, a mathematician and economist by training, is with the National Accounts Directorate at Eurostat (the Statistical Office of the European Union). He played a central role in the revision of the international System of National Accounts. He now works on developing environmental accounting in Europe.

HENRY M. PESKIN is President of Edgevale Associates, a consulting company. He was formerly on the staffs of the National Bureau of Economic Research, the Urban Institute, and the Institute for Defense Analysis. Most recently, he was a senior fellow at Resources for the Future. With training in chemical engineering, an undergraduate degree in political science, and a graduate degree in economics, he has written extensively on methods to expand the national economic accounts in order to better measure resource and environmental degradation. As a consultant to the World Bank and the U.S. Agency for International Development, he has surveyed environmental accounting practices in industrialized countries and has advised developing countries on the design and implementation of environmental accounting systems.

JOHN REILLY is Associate Director for the Joint Program on Science and Policy on Global Change at the Massachusetts Institute of Technology energy laboratory. He was Deputy Director of the Natural Resource and Environment Division of the U.S. Department of Agriculture's Economic Research Service during the writing of this report. He holds a Ph.D. in economics from the University of Pennsylvania. He has written numerous articles, chapters, and books in the areas of climate change, agriculture, and natural-resource economics.

ROBERT REPETTO is Tim Wirth Fellow in the Graduate School of Public Affairs at the University of Colorado at Denver. He is also an advisor at Stratus Environmental Consulting, Inc., in Boulder, Colorado. Before moving to Boulder, he was Vice President of the World Resources Institute (WRI), a nonprofit policy research center in Washington, D.C., and he remains affiliated with WRI as Senior Research Fellow. He is known for his writings and research on the interface between environment and economics and on measures to promote sustainable economic development. He is currently a member of the National Research Council's Board on Sustainable Development (Population Division) and recently completed a 3-year term on the Environmental Economics Advisory Committee of the Environmental Protection Agency's Science Advisory Board. He has been awarded a Pew Fellowship in Marine Conservation for the period 1998-2000. He has been a Professor at the Harvard School of Public Health, a World Bank official working in Indonesia, an economic advisor in Pakistan, a Ford Foundation staff economist in India, and an economic analyst at the Federal Reserve Bank of New York.

BRIAN J. SKINNER is a Professor of Geology and Geophysics at Yale University. He has served on many National Research Council committees. He has served as Cochair of the Board on Earth Sciences and Resources (Commission on Geosciences, Environment, and Resources); chair of the Board on Earth Sciences (Commission on Physical Sciences, Mathematics, and Resources), the Committee on Mineral Resources and the Environment (Commission on Natural Resources), and the U.S. National Committee for the International Union of Geological Sciences (Commission on Geosciences, Environment, and Resources); and member of the Board on Mineral and Energy Resources (Commission on Physical Sciences, Mathematics, and Applications). He has also been an ex-officio member of the U.S. National Committee for the International Geographical Union, the U.S. Committee for Geochemistry, the U.S. Committee on the History of Geology, the U.S. National Committee for the International Geological Correlation Program, the U.S. Committee for the International Association of Engineering Geology, the U.S. Committee for the International Association of Hydrogeologists, and the U.S. Committee for the International Association for Mathematical Geology (all Commission on Geosciences, Environment, and Resources).

JOHN E. TILTON is William J. Coulter Professor of Mineral Economics at the Colorado School of Mines. He is also a university fellow at Resources for the Future and a past president of the Mineral Economics and Management Society. His teaching and research interests over the past 25 years have focused on economic and public policy issues associated with the mineral industries. He has served as Vice-Chair of the Board on Mineral and Energy Resources and as a member of a number of other National Research Council boards and committees.

VICTORIA J. TSCHINKEL, a zoologist by training, is Senior Consultant for Strategic Environmental Management with Landers and Parsons, P.A. She served as Secretary of the Florida Department of Environmental Regulation from 1981 to 1987. She is currently a member of the Commission on Geosciences, Environment, and Resources; a corporate director for a major petroleum company; a director of the German Marshall Fund; a director of Resources for the Future; and a member of the Board of the University of Chicago Governors of Argonne National Laboratory. She is a director of 1000 Friends of Florida and Florida Communities Trust. She has served on numerous committees of the National Research Council on a wide range of topics.

MARTIN L. WEITZMAN is Ernest E. Monrad Professor of Economics at Harvard University. Previously, he was Mitsui Professor of Economics at the Massachusetts Institute of Technology. He has been elected a fellow of the American Academy of Arts and Sciences and of the Econometric Society. He has written extensively on environmental economics and on biodiversity and was awarded the prize for publication of enduring merit from the Association for Environmental and Resource Economics.

References

Adelman, M.A.
  1990   Mineral depletion, with special reference to petroleum. *Review of Economics and Statistics* 72(1):1-10.
Adelman, M.A., and G.C. Watkins
  1996   *The Value of United States Oil and Gas Reserves*. MIT-CEEPR 96-004 WP. Cambridge, MA: Massachusetts Institute of Technology.
Advisory Commission to Study the Consumer Price Index
  1996   *Toward a More Accurate Measure of the Cost of Living: The Final Report of the Advisory Commission to Study the Consumer Price Index*. Washington, DC: Advisory Commission to Study the Consumer Price Index.
Anderson, R., and M. Rockel
  1991   Economic Valuation of Wetlands. Discussion Paper #65, American Petroleum Institute. Washington, DC.
Aronsson, T., P.-O. Johansson, and K.-G. Löfgren
  1997   *Welfare Measurement, Sustainability and Green National Accounting*. Cheltenham, U.K.: Edward Elgar.
Arrow, K., R. Solow, P. Portney, E. Leamer, R. Radner, and H. Schuman
  1993   Report of the National Oceanographic and Atmospheric Administration Panel on Contingent Valuation. *Federal Register* (January) 58:4601-4614.
Ayres, R.U., and A.V. Kneese
  1969   Production, consumption, and externalities. *American Economic Review* LIX(3): 282-297.
Bartelmus, P.
  1994   *Environment, Growth and Development—The Concepts and Strategies of Sustainability*. London and New York: Routledge.
  1998   The value of nature-valuation and evaluation in environmental accounting. In Uno and Bartelmus, eds., *Environmental Accounting in Theory and Practice*. Kluwer: Dordrecht.

Berck, P.
1979   The economics of timber: A renewable resource in the long run. *Bell Journal of Economics* 10:447-462.

Bergstrom, J.C., J.M. Bowker, H.K. Cordell, G. Bhat, D.B.K. English, R.J. Teasley, and P. Villegas
1996   Ecoregional Estimates of the Net Economic Value of Outdoor Recreational Activities in the United States: Individual Model Results. Final report. U.S. Department of Agriculture, Forest Service, Washington, DC.

Binkley, C.S., and J.R. Vincent
1988   Timber prices in the U.S. South: Past trends and outlook for the future. *Southern Journal of Applied Forestry* 12:15-18.

Birdsey, R.A., and L.S. Heath
1995   Carbon changes in U.S. forests. 70 pp. in L. Joyce, ed., *Productivity of American Forests and Climate Change.* General Technical Report RM-271, USDA FS Pub., Rocky Mountain Forest and Range Experimental Station, Fort Collins, CO.

Bowes, M., J. Krutilla, and T. Stockton
1984   Forest management for increased timber and water yields. *Water Resources Research* 20:655-663.

Braden, J.B., and C.D. Kolstad, eds.
1991   *Measuring the Demand for Environmental Quality.* Amsterdam: North-Holland, Elsevier Publishers B.V.

Brown, S.
1996   Managing forests for mitigation of greenhouse gases. Pp. 773-798 in R.T. Watson, M.C. Zinyowera, and R.M. Moss, eds., *Climate Change 1995: Impacts, Adaptations, and Mitigation of Climate: Scientific and Technical Analyses.* Published for the Intergovernmental Panel on Climate Change. Cambridge: Cambridge University Press.

Bruce, J.P., H. Lee, and E.H. Haites, eds.
1996   *Climate Change 1995: Economic and Social Dimensions.* Published for the Intergovernmental Panel on Climate Change. Cambridge, U.K.: Cambridge University Press.

Bureau of Economic Analysis
1982   Measuring Nonmarket Economic Activity. BEA Working Paper No. 2, December.
1987   Environmental Economics Division History. Unpublished manuscript, February 6.
1994a   Integrated economic and environmental satellite accounts. *Survey of Current Business* April:33-49.
1994b   Accounting for mineral resources. *Survey of Current Business* April:50-72.
1995a   *Mid-Decade Strategic Review of BEA's Economic Accounts: Background Papers.* Washington, DC: U.S. Department of Commerce.
1995b   *An Introduction to National Economic Accounting.* Methodology Paper Series MP-1. Washington, DC: U.S. Department of Commerce.

Bureau of Mines
1972   *Minerals Yearbook 1970, Vol. II.* Washington, DC: U.S. Department of the Interior.
1987   *An Appraisal of Minerals Availability for 34 Commodities.* Washington, DC: U.S. Department of the Interior.

Cairns, R.D.
1997   Accounting for Resource Depletion. Unpublished paper, McGill University.

Cairns, R.D., and G.A. Davis
    1998a   Valuing petroleum reserves using current net price. *Economic Inquiry* (forthcoming).
    1998b   On using current information to value hard-rock mineral properties. *Review of Economics and Statistics* (forthcoming).

Carson, C.S.
    1975   The history of the United States National Income and Product Accounts: The development of an analytical tool. *Review of Income and Wealth* June:153-181.

Carson, R., and R. Mitchell
    1993   The value of clean water: The public's willingness to pay for boatable, fishable, and swimmable quality water. *Water Resources Research* 29(7):2445-2454.

Christensen, L.R., and D.W. Jorgenson
    1969   The measurement of U.S. real capital input, 1929-1967. *Review of Income Wealth* 15(4):293-320.
    1973   Measuring economic performance in the private sector. Pp. 233-338 in *The Measurement of Economic and Social Performance.* Studies in Income and Wealth, Vol. 38. New York: Columbia University Press for the National Bureau of Economic Research.

Clawson, M.
    1979   Forests in the long sweep of American history. *Science* 204:1168-1174.

Clean Air Act Council on Compliance
    1997   Report of the Clean Air Act Council on Compliance. Letter report to the EPA Administrator, private communication.

Commission of the European Communities
    1993   Commission of the European Communities, International Monetary Fund, Organization for Economic Cooperation and Development, United Nations, and World Bank, Brussels and elsewhere. *System of National Accounts.*

Cornes, R., and T. Sandler
    1986   *The Theory of Externalities, Public Goods, and Club Goods.* New York: Cambridge University Press.

Costanza, R., R. d'Arge, R. de Groot, F. Farber, M. Grasso, B. Hannon, K. Limburg, S. Naeem, R.V. O'Neill, J. Pervello, R.Q. Raskin, P. Sutton, and M. Van den Bélt
    1997   The value of the world's ecosystem services and natural capital. *Nature* (May) 387:253-260.

Council of Economic Advisers
    1995   *Economic Report of the President,* Council of Economic Advisers, pp. 216-219. Washington, DC: U.S. Government Printing Office.

Craig, J.R., D.J. Vaughan, and B.J. Skinner
    1988   *Resources of the Earth.* Englewood Cliff, NJ: Prentice-Hall.

Crowards, T.M.
    1996   Natural resource accounting: A case study of Zimbabwe. *Environmental and Resource Economics* 7(2):213-241.

Daly, H.E., and J.B. Cobb, Jr.
    1989   *For the Common Good: Redirecting the Economy Toward Community, The Environment, and a Sustainable Future.* Boston: Beacon Press.
    1994   *For the Common Good: Redirecting the Economy Toward Community, The Environment, and a Sustainable Future.* Second edition. Boston: Beacon Press.

Davis, G.A.
    1996   Option premiums in mineral asset pricing: Are they important? *Land Economics* 72:167-186.

1997 Valuing the Stock and Flow of Mineral and Renewable Assets in National Income Accounting. Final Technical Report, NSF/EPA Partnership for Environmental Research, Grant No. R824705-01-0.

Davis, G.A., and D. Moore
1997 Valuing Mineral Stocks and Flows in Environmental-Economic Accounts. Unpublished paper, Colorado School of Mines.
1998 Valuing Mineral Reserves When Capacity Constrains Production. *Economics Letters* 60(1):121-125.

de Boo, A.J., P.R. Bosch, C.N. Gorter, and S.J. Keuning
1991 An environmental module and the complete system of national accounts. Number NA-046. Netherlands Central Bureau of Statistics.

Denison, E.F.
1979 Pollution abatement programs: Estimates of their effect upon output per unit of input, 1975-78. *Survey of Current Business* 59(8, part I):58-59.

Diamond, P.A., and J.A. Hausman
1994 Contingent valuation: Is some number better than no number? *Journal of Economic Perspectives* 8(4):45-64.

Duchin, F., and G. Lange
1993 Development and the Environment in Indonesia: An Input-Output Analysis of Natural Resource Issues. Final report to the Canadian International Development Agency, August.

Eisner, R.
1971 New twists to income and product. *Survey of Current Business Part II* (Anniversary Issue—*The Economic Accounts of the United States: Retrospect and Prospect*) 51(7):67-68.
1985 The total incomes system of accounts. *Survey of Current Business* 65(1):24-28.
1988 Extended accounts for national income and product. *Journal of Economic Literature* (December) 26:1611-1684.
1989 *The Total Incomes System of Accounts.* Chicago: University of Chicago Press.

El Serafy, S.
1989 The proper calculation of income for depletable natural resources. Pp. 10-18 in Y.J. Ahmad, S. El Serafy, and E. Lutz, eds., *Environmental Accounting for Sustainable Development.* Washington, DC: The World Bank.

European Union
1994 *Directions for the European Union on Environmental Indicators and Green National Accounting: The Integration of Environmental and Economic Accounting Systems.* Reference COM(94)670. Luxembourg: Office for Official Publications of the European Communities.

Fisher, A.C.
1981 *Resource and Environmental Economics.* Cambridge, U.K.: Cambridge University Press.

Fraumeni, B.M.
1997 The measurement of depreciation in the U.S. National Income and Product Accounts. *Survey of Current Business* (July):7-23.

Gianessi, L.P., and H.M. Peskin
1976 The Costs to Industries of Meeting the 1977 Provisions of the Water Pollution Control Act Amendments of 1972. Report to the U.S. Environmental Protection Agency, January.

Grambsch, A., and R.G. Michaels
1994 The United Nations Integrated Environmental and Economic Accounting System: Is It Right for the U.S.? Paper presented to the 1994 AERE Workshop Integrating the Environment and the Economy: Sustainable Development and Economic/Ecological Modeling, June 6. Boulder, CO.

Green, I.M., C. Folke, K. Turner, and I. Mateman
  1994   Primary and secondary valuation of wetland ecosystems. *Environmental and Resource Economics* 4:55-74.
Hanemann, W.M.
  1994   Valuing the environment through contingent valuation. *Journal of Economic Perspectives* 8(4):19-43.
Hartwick, J.M., and A. Hageman
  1993   Economic depreciation of mineral stocks and the contribution of El Serafy. In Ernst Lutz, ed., *Toward Improved Accounting for the Environment*. Washington, DC: The World Bank.
Hay, M.J.
  1988   Net Economic Recreation Value for Deer, Elk, and Waterfowl Hunting and Bass Fishing. Report 85-1, U.S. Department of the Interior, U.S. Fish and Wildlife Service.
Hicks, J.R.
  1939   *Value and Capital*, 2nd ed. Oxford, U.K.: Clarendon Press.
  1940   The valuation of the social income. *Economica* 7(2):105-124.
Hoehn, J.P., and J.B. Loomis
  1993   Substitution Effects in the Valuation of Multiple Environmental Programs. *Journal of Environmental Economics and Management* 25:56-75.
Howell, S.L.
  1996   A review of the conceptual and methodological issues in accounting for forests. In *Proceedings of Third Meeting of the London Group on Natural Resource and Environmental Accounting*. Stockholm: Statistics Sweden.
Japan, Economic Council of
  1973   *Measuring Net National Welfare of Japan*. NNW Measurement Committee. Tokyo: Printing Bureau, Ministry of Finance.
Jensen, H.V., and O.G. Pedersen
  1998   Danish NAMEA 1980-1992. Copenhagen: Statistics Denmark.
Jordan, J., and A. Elnagheeb
  1993   Willingness to pay for improvements in drinking water quality. *Water Resources Research* 29:237-245.
Jorgenson, D.W., and B.M. Fraumeni
  1987   The accumulation of human and non-human capital, 1948-1984. Unpublished manuscript, Harvard University.
Jorgenson, D.W., and P.J. Wilcoxen
  1990   Environmental regulation and U.S. economic growth. *Rand Journal of Economics* Summer 21(2):314-340.
Juster, F.T.
  1973   A framework for the measurement of economic and social performance. In M. Moss, ed., *The Measurement of Economic and Social Performance, Studies in Income and Wealth*, vol. 38. New York: Columbia University Press for the National Bureau of Economic Research.
Kendrick, J.W.
  1987   Happiness is personal productivity growth. *Challenge* 30(2):37-44.
  1996   *The New System of National Accounts*. Boston: Kluwer Academic Publishers.
Keuning, S.J.
  1993   An information system for environmental indicators in relation to the national accounts. In W.F.M. deVries, G.P. den Bakker, M.G.B. Gircour, S.J. Keuning, and A. Lenson, eds., *The Value Added of National Accounting*. Voorburg/Heerlen: Netherlands Central Bureau of Statistics.

Kilburn, L.C.
   1990   Valuation of mineral properties which do not contain exploitable reserves. *CIM Bulletin* 83(940):90-93.
Kuznets, S.
   1948a   On the valuation of social income—Reflections on Professor Hicks' article, part I. *Economica* XV(57):1-16.
   1948b   On the valuation of social income—Reflections on Professor Hicks' article, part II. *Economica* XV(58):116-131.
   1948c   National income: A new version (Discussion of the new Department of Commerce Income Series). *Review of Economics and Statistics* 30(3):151-179.
Lant, C.L., and R.S. Roberts
   1990   Greenbelts in the cornbelt: Riparian wetlands, intrinsic values, and market failure. *Environment and Planning* 22:1375-1388.
Leontief, W.
   1970   Environmental repercussions and the economic structure: An input-output approach. *The Review of Economics and Statistics* 52(3):262-271.
Lind, R.C.
   1990   Reassessing the government's discount rate policy in light of new theory and data in a world economy with a high degree of capital mobility. *Journal of Environmental Economics and Management* 18(2):S8-S28.
   1997   Intertemporal equity, discounting, and economic efficiency in water policy valuation. *Climate Change* 37(1):41-62.
Loomis, J.B., and D.S. White
   1996   Economic benefits of rare and endangered species: Summary and meta-analysis. *Ecological Economics* 18:197-206.
Lyon, K.S.
   1981   Mining of the forest and the time path of the price of timber. *Journal of Environmental Economics and Management* 8:330-344.
McCollum, D.W., G.L. Peterson, J.R. Arnold, D.C. Markstrom, and D.M. Hellerstein
   1990   The Net Economic Value of Recreation on the National Forests: Twelve Types of Primary Activity Trips Across Nine Forest Service Regions. Research Paper RM-289, February. U.S. Department of Agriculture, Forest Service, Rocky Mountain Forest and Range Experiment Station, Fort Collins, CO.
Miller, M.H., and C.W. Upton
   1985   A test of the Hotelling valuation principle. *Journal of Political Economy* 93:1-25.
Mitchell, R.C., and R.T. Carson
   1989   *Using Surveys To Value Public Goods: The Contingent Valuation Method.* Washington, D.C.: Resources for the Future.
National Research Council
   1997   *Valuing Groundwater.* Committee on Valuing Ground Water, Commission on Geosciences, Environment, and Resources. Washington, DC: National Academy Press.
Nordhaus, W.D., and D. Popp
   1997   What is the value of scientific knowledge? An application to global warming using the PRICE model. *The Energy Journal* 18(1):1-45.
Nordhaus, W.D., and J. Tobin
   1972   Is growth obsolete? In *Economic Growth*, 50th anniversary colloquium V. New York: Columbia University Press for the National Bureau of Economic Research; also published in M. Moss, ed., *The Measurement of Economic and Social Performance, Studies in Income and Wealth*, vol. 38, 1973. New York: Columbia University Press for the National Bureau of Economic Research.
Okun, A.M.
   1971   Social welfare has no price tag. *Survey of Current Business* 51(7):129-33.

Parker, R.P.
  1991   A Preview of the Comprehensive Revisions of the National Income and Product
          Accounts. *Survey of Current Business* 71(10):20-28.
  1996   The International System of National Accounts (SNA). Presentation by R. Parker to
          the Panel on Integrated Environmental and Economic Accounting, October 17.
          Washington, DC.
Pearce, D.W., W.R. Cline, A.N. Achanta, S. Fankhauser, R.K. Pachauri, R.S.J. Tol, and P.
Vellinga
  1996   The social costs of climate change: Greenhouse damage and the benefits of con-
          trol. Pp. 145-178 in J.P. Bruce, H. Lee, and E.H. Haites, eds., *Climate Change 1995:
          Economic and Social Dimensions.* Cambridge, U.K.: Cambridge University Press
          for the Intergovernmental Panel on Climate Change.
Perrings, C.I.
  1998   Resilience in the dynamics of economy-environment systems. *Environmental and
          Resource Economics* 11(3-4): 503-520.
Peskin, H.M.
  1989a   *Accounting for Natural Resource Depletion and Degradation in Developing Countries.*
          Environmental Department Working Paper No. 13. Washington, DC: The World
          Bank.
  1989b   A proposed environmental accounts framework. Pp. 65-78 in Y.J. Ahmad, S. El
          Serafy, and E. Lutz, eds., *Environmental Accounting for Sustainable Development.*
          Washington, DC. The World Bank.
Phillips, W.E., T.J. Haney, and W.L. Adamowicz
  1993   An economic analysis of wildlife habitat preservation in Alberta. *Canadian Jour-
          nal of Agricultural Economics* 41:411-418.
Portney, P.R.
  1994   The contingent valuation debate: Why economists should care. *Journal of Eco-
          nomic Perspectives* 8(4):3-17.
Portney, P.R., and J.P. Weyant
  1999   *Discounting and Intergenerational Equity.* Washington, DC: RFF Press.
Randall, A., and J. Stoll
  1983   Existence value in a total valuation framework. In R. Rowe and L. Chestnut, eds.,
          Managing Air Quality and Scenic Resources at National Parks and Wilderness Areas.
          Boulder, CO: Westview Press.
Reilly, J., and K. Richards
  1993   An economic interpretation of the trace gas index issue. *Environmental and Re-
          source Economics* 3:41-61.
Repetto, R.
  1986   *World Enough and Time.* New Haven, CT: Yale University Press.
Repetto, R., W. Magrath, M. Wells, C. Beer, and F. Rossini
  1989   *Wasting Assets: Natural Resources in the National Income Accounts.* Washington,
          DC: World Resources Institute.
Ribaudo, M.O.
  1989   Water quality benefits from the conservation reserve program. *Agricultural Eco-
          nomics,* Report No. 606. Washington, DC: U.S. Department of Agriculture, Eco-
          nomic Research Service.
Ribaudo, M.O., and S. Piper
  1991   Estimating changes in recreational fishing participation from national water qual-
          ity policies. *Water Resources Research* 27(7):1757-63.
Samuelson, P.A.
  1954   The pure theory of public goods. *The Review of Economics and Statistics* 36(4):387-
          389.

1955   Diagrammatic exposition of a theory of public expenditure. *The Review of Economics and Statistics* 37(4):350-356.

Schelling, T.C.
1995   Intergenerational discounting. *Energy Policy* 23:395-401.

Sedjo, R.
1990   The Nation's Forest Resources. Discussion Paper ENR 90-07. Washington, DC: Resources for the Future.

Sedjo, R.A., and K.S. Lyon
1990   *The Long-Term Adequacy of World Timber Supply.* Washington, DC: Johns Hopkins University Press.

Smil, V., and M. Yshi
1998   *The Economic Costs of China's Environmental Degradation.* Cambridge, MA: American Academy of Arts and Sciences.

Smith, V.K.
1993   Nonmarket valuation of environmental resources: An interpretive appraisal. *Land Economics* 69(1):1-26.
1996   *Estimating Economic Values for Nature: Methods for Non-Market Valuation.* Northamton: Edward Elgar.

Solow, R.
1992   An Almost Practical Step Toward Sustainability. Resources for the Future Invited Lecture, Washington, DC, October 8.

Sorg, C.F., and J.B. Loomis
1984   Empirical Estimates of Amenity Forest Values: A Comparative Review. General Technical Report RM-107, March. U.S. Department of Agriculture, Forest Service, Rocky Mountain Forest and Range Experiment Station. Fort Collins, CO.

Sun, H., J. Bergstrom, and J. Dorfman
1992   Estimating the benefits of groundwater contamination control. *Southern Journal of Agricultural Economics* 24:63-71.

Survey of Current Business
1994   Accounting for mineral resources: Issues and BEA's initial estimates. *Survey of Current Business* April:50-72.

Torries, T.F.
1988   Competitive cost analysis in the mineral industries: The example of nickel. *Resources Policy* September:193-204.
1995   Comparative costs of nickel sulphides and laterites. *Resources Policy* 21(3):179-187.

United Nations
1984   A Framework for the Development of Environment Statistics. Sales No. E.84.XVII.12.
1991   Concepts and Methods of Environment Statistics: Statistics of the Natural Environment—A Technical Report. Sales No. E.91.XVII.18.
1992a   Framework Convention on Climate Change, Report of the Intergovernmental Negotiating Committee for a Framework Convention on Climate Change, Part II, Add. 1, Corr. 1.
1992b   Convention of Biological Diversity, Environmental Law and Institutions Programme Activity Centre, June.
1993   *Integrated Environmental and Economic Accounting: Interim Version, Studies in Method,* Series F, Number 61. New York: The United Nations.

Uno, K., and P. Bartelmus
1998   Environmental Accounting in Theory and Practice. Dordrecht, Boston, and London: Kluwer Academic Publishers.

U.S. Congress
  1995   House Report Accompanying HR4603, U.S. Department of Commerce, FY 1995,
          Public Law 103-317. Washington, DC.
U.S. Department of Agriculture Forest Service
  1995   Chapter 4, Socioeconomic Effects and Implications of the Proposed Program, 1995
          draft RPA program. http://www.fs.fed.us/land/RPA/chapt4.htm.
U.S. Department of Commerce
  1954   *National Income: 1954 Edition.* Washington, DC: U.S. Government Printing Of-
          fice.
U.S. Environmental Protection Agency
  1996   National Air Quality and Emissions Trend Report, 1995. EPA 454-R-96-015 Of-
          fice of Air Quality Planning and Standards, Research Triangle Park, NC, October.
  1997   The Benefits and Costs of the Clean Air Act, 1970 to 1990. Draft, April. Office of
          Air and Radiation/Office of Policy Analysis and Review/Office of Policy, Plan-
          ning, and Evaluation.
U.S. Geological Survey
  1992   Mineral commodities summaries, p. 203. Washington, DC: Bureau of Mines,
          U.S. Department of the Interior.
Vincent, J.
  1997   Net Accumulation of Timber Resources. Manuscript, Harvard Institute for Inter-
          national Development, Cambridge, MA.
Vincent, J., and J.M. Hartwick
  1997   Accounting for the Benefits of Forest Resources: Concepts and Experience. Draft,
          July 10. FAO Forestry Department.
Waddington, D.G., K.J. Boyle, and J. Cooper
  1994   1991 Net Economic Values for Bass and Trout Fishing, Deer Hunting, and Wild-
          life Watching. Report 91-1, October. U.S. Department of the Interior, U.S. Fish
          and Wildlife Service.
Walsh, R.G., D.M. Johnson, and J.R. McKean
  1988   Review of Outdoor Recreation Economic Demand Studies with Nonmarket Ben-
          efit Estimates, 1968-1988. Technical Report No. 54, December. Colorado Water
          Research Institute, Colorado State University, Fort Collins, CO.
Washburn, C.L.
  1990   The Determinants of Forest Value in the U.S. South. Ph.D. thesis, Yale Univer-
          sity.
Weitzman, M.
  1976   On the welfare significance of national product in a dynamic economy. *Quarterly
          Journal of Economics* 90:156-162.
World Bank
  1997   *Expanding the Measure of Wealth: Indicators of Environmentally Sustainable Develop-
          ment.* Washington, DC: The World Bank.
World Commission on Environment and Development
  1987   *Our Common Future.* Oxford, U.K.: Oxford University Press.
Young, A.H., and H.S. Tice
  1985   An Introduction to National Economic Accounting. *Survey of Current Business*
          65(3):59-74, 76.
Young, C.E.F., and R. Seroa da Motta
  1995   Measuring Sustainable Income from Mineral Extraction in Brazil. *Resource Policy*
          21(2):113-125.
Zolotas, X.
  1981   *Economic growth and declining social welfare.* Athens: Bank of Greece.

Index

A

Abatement, *see* Pollution abatement and control

Accounts and accounting, definition and purpose, 12, 19, 27, 207

Acid precipitation, 40, 113, 116, 119, 121, 122, 142, 143, 145, 147, 148

Aggregation and disaggregation, 14, 19, 32, 36, 37, 39-40, 84, 123, 152, 186
 see also Gross domestic product; Gross national product

Agriculture, 34, 115, 168, 199
 air pollutants, 114, 144, 145, 146, 147
 forests and, 132, 133, 135, 138-139
 hedonic valuation, 118
 IEESA, 3, 106, 109, 111-112, 118, 168, 169
 land and structures valuation, 111-112, 168
 production costs, 135
 see also Department of Agriculture

Air quality and pollution, 8, 32, 113-116, 121-122, 126, 141-149, 152, 178-180, 199
 acid precipitation, 40, 113, 116, 119, 121, 122, 142, 143, 145, 147, 148
 agriculture, impacts on, 114, 144, 145, 146, 147
 BEA methodology, 148-149
 carbon dioxide, 134, 135, 136, 176, 208

carbon monoxide, 142-143, 145, 208-209

chlorofluorocarbons, 143

data requirements/availability, 149, 171-172, 199

defined, 207

ecosystem impacts, 149, 156

forests, 34, 145, 147
 climate change and, 34, 134, 135, 136, 158, 178

fossil fuels, general, 60; *see also* Climate change

historical perspectives, 142, 145, 146, 147, 148

IEESA, 3, 55, 110, 112, 156-157, 160, 171-172

market forces, 142, 144-145

nitrogen dioxide, 143, 144, 145

ozone, ground-level, 143, 144, 145, 147

ozone layer depletion, 23, 115, 121, 122, 143-144, 145, 149, 218

particulate matter, 143, 144, 145, 146, 147, 218

photochemical, 218

recreation and, 114, 149, 156

residuals, 7, 113-116, 119, 121, 147, 150

secondary, 220

SEEA, 112

sulfur dioxide, 143, 144, 145; *see also* "acid precipitation" supra

239

H

I